7 v15 (19)

Wetware

Wetware

A Computer in Every Living Cell

Dennis Bray

Yale University Press   New Haven & London

Published with assistance from the foundation
established in memory of Philip Hamilton McMillan of
the Class of 1894, Yale College.

Set in Bulmer Roman by Binghamton Valley Composition.
Printed in the United States of America.

Library of Congress Cataloging-in-Publication Data

Bray, Dennis.
Wetware : a computer in every living cell / Dennis Bray.
p.   cm.
Includes bibliographical references and index.
ISBN 978-0-300-14173-3 (cloth : alk. paper) 1. Cytology. 2. Cell interaction. 3. Molecular
biology. 4. Computational biology. I. Title.

QH581.2.B736 2009
571.6—dc22   2008044720

A catalogue record for this book is available from the British Library.

This paper meets the requirements of ANSI/NISO Z39.48-1992 (Permanence of Paper).
It contains 30 percent postconsumer waste (PCW) and is certified by
the Forest Stewardship Council (FSC).

10   9   8   7   6   5   4   3   2   1

For Maggie, Claire, Simon, and Phoebe, with love

A goal for the future would be to determine the extent of knowledge the cell has of itself and how it utilizes this knowledge in a "thoughtful" manner when challenged.

—Barbara McClintock, Nobel Prize acceptance lecture, 1983

Contents

Preface

There is a real risk in writing this book of being misunderstood. One of the rejection slips I received, after sending the manuscript to a large publishing house, asserted that it was about single-celled organisms possessing consciousness. Not true! I say repeatedly in the book as clearly as English words will allow that in my opinion single cells are not sentient or aware in the same way that we are. To me, consciousness implies intelligent awareness of self and the ability to experience introspectively accessible mental states. No single-celled organism or individual cell from a plant or animal has these properties. An individual cell, in my view, is a system that possesses the basic ingredients of life but lacks sentience. It is a robot made of biological materials.

It cannot be denied, however, that those systems that do possess consciousness—principally human beings—are themselves made of cells. A very large number of cells, it is true, and linked in highly complex ways, but cells for all that. Moreover, there is a direct link in evolution and development between a single cell and humans. Cells are undeniably the "stuff" from which consciousness is made.

Some say that organization is paramount. If we were able to replace each nerve cell in our brain with an equivalent silicon device, they claim, then the outcome would be an entity with all the mental states of the original. The idea that computers of the future will be sentient and experience internal mental states is the starting point of many science fiction stories, part of the zeitgeist. But this is a theory without evidence. We do not know it to be true. My own view, as you will see, is that present-day electronic devices and robots are woefully inadequate

ix

in this regard. They lack the multiplicity of states and plasticity displayed by living systems; they are unable to construct and repair themselves.

Living cells have an unlimited capacity to detect and respond to their surroundings. An unending kaleidoscope of environmental challenges has been present throughout evolution. Organisms have responded by changing their chemistry; any that failed to adjust became extinct. And the richest source of variation was in the giant molecules that distinguish living systems. From a time-compressed view, the sequences and structures of RNA, DNA, and proteins can be thought of as continually morphing in response to the fluctuating world around them. These changes are cumulative with each modification adding to those that have gone before. It is as though each organism builds an image of the world—a description expressed not in words or in pixels but in the language of chemistry. Every cell in your body carries with it an abstraction of its local surroundings in constellations of atoms. A basic knowledge of and response to the environment are integral parts of every living cell's makeup.

The term *wetware* is not new, but I think it has not been closely defined before. Wetware, in this book, is the sum of all the information-rich molecular processes inside a living cell. It has resonance with the rigid *hardware* of electronic devices and the symbolic *software* that encodes memories and operating instructions, but is distinct from both of these. Cells are built of molecules that interact in complex webs, or circuits. These circuits perform logical operations that are analogous in many ways to electronic devices but have unique properties. The computational units of life—the transistors, if you will—are its giant molecules, especially proteins. Acting like miniature switches, they guide the biochemical processes of a cell this way or that. Linked into huge networks they form the basis of all of the distinctive properties of living systems. Molecular computations underlie the sophisticated decision making of single-cell organisms such as bacteria and amoebae. Protein complexes associated with DNA act like microchips to switch genes on and off in different cells—executing "programs" of development. Machines made of protein molecules are the basis for the contractions of our muscles and the excitable, memory-encoding plasticity of the human brain. They are the seed corn of our awareness and sense of self.

When a friend asked me who this book was for, I ingenuously answered, "Myself." Over the years I had acquired a ragbag of unanswered questions relating to living systems, computers, and consciousness and it was time to think them through and put them into order. So I did indeed set out, as John Steinbeck says in his *Travels with Charley,* "not to instruct others but to inform myself." But the discipline of writing calls for a voice and demands an imaginary reader. As I worked I found myself laying out my arguments as clearly as possible to someone lacking specialized background in biology or computers. My imaginary reader has a high school or equivalent background in basic science and a philosophical inclination. Ideally, she is already interested in such things as the comparison of living systems and computers and the origins of sentient properties from inanimate matter.

The central thesis of the book—that living cells perform computations—arises from contemporary findings in the biological sciences, especially biochemistry and molecular biology. It is a leitmotif of systems biology, although the philosophical ramifications of that new discipline are rarely expressed. Many readers with direct experience of computer-based games and virtual environments will also have wondered about their relationship to the world of real organisms. I hope that they will find here an elaboration if not an answer to their questions.

This book took shape over many years and owes much to friends and colleagues. Hamid Bolouri and Armand Leroi saw an early version, and I am grateful for their positive response despite obvious flaws. Graeme Mitchison read the manuscript from beginning to end, and his comments took the book to a higher level. At a later stage, Horace Barlow made crucial improvements to the text as well as adding his considerable insight into the way the brain works. Aldo Faisal, Steve Grand, Frank Harold, Dan Heaton, Auke Ijspeert, Lizzie Jeffries, Dale Purves, Hugh Robinson, John Scholes, Yuhai Tu, Rob White, Bé Wieringa, and Alan Winfield each helped me in difficult areas and made valuable suggestions. Claire Strom, super editor, went through the text like a butcher with a cleaver, flensing away the pompous verbiage we scientists are so fond of. Her daughter, Phoebe, age fifteen, used a lighter touch to identify missing explanations ("Sometimes I think I get this and then it goes Poof!"). Literary agent Peter Tallack and Yale editor Jean Thomson Black combined professional criticism with a genuine enthusiasm for the project that carried me along. Thank you all.

ONE

Clever Cells

It was a rainy November Cambridge afternoon when Bill Grimstone appeared at my office in the Zoology Department and said he had something to show me. It was rare, even during the term, to sight him, and most unusual for him to be in such an animated state. Bill was an archetypal imperturbable Cambridge don: suave, phlegmatic, with graying hair, spectacles and a slight cast in one eye, and given to wearing a tweed jacket and a tie. As I followed him down the corridor to his room, I speculated that there could be only one reason for this excitement—his research. Sure enough, as he ushered me into his small office, he gestured toward a wooden chair in front of a microscope. Even before he flicked the switch to activate the light, I knew I would be looking at termite guts.

Termites live by eating and digesting wood. In the tropics they build huge colonies like pillars, and, I gather, they can be serious pests if they settle into your home. I've also learned that termites, to gain nourishment from wood, have to degrade wood's primary component, cellulose, and that this requirement presents a biochemical challenge. Cellulose is just a chain of glucose subunits. But animals cannot digest this potentially rich source of food, for reasons that have always been a mystery to me. You might have thought that an evolving organism would easily acquire the single enzyme (a protein performing a specific reaction) needed to tap into such a potentially rich source of energy. But the fact is that any animal, including an insect, that wants to digest wood

1

must recruit bacteria. Termites do so by turning the gut into an oxygen-free chamber full of special bacteria that degrade cellulose: a mutually beneficial ménage because the termite provides the bacteria with a constant supply of well-chewed wood fragments to digest. In return the bacteria turn the wood into sugars and other easily digestible molecules. They take some of the nutrients for their own use and leave the rest for their insect host.

So as I looked down Bill's microscope I saw, as expected, a jumble of wood fragments surrounded by the dark forms of bacteria, rounded or rod-shaped. But as I fumbled with the unfamiliar controls, something altogether more formidable slid into view. It was a single cell, but as unlike the textbook fried-egg image of a cell as one could imagine. This was a huge Wurlitzer of a cell, covered from head to foot with writhing snakelike flagella—protrusions cells use to drive them through water. Every portion of its body, which seemed immense under the powerful magnification of the microscope, moved with its own rhythm, as though driven by cogs and machines beneath the carapace. As I passed the eyepiece to Bill, the writhing circular motion continued, unfazed by our observation. "*Trichonympha,*" Bill explained in his cultured baritone. "And here," as he searched with the microscope stage, "is *Streblomastix,* with a background of *Spirochaetes.*" He had left the microscope focused on a large serpentine body that bristled with surface hairs surrounded by darting helical structures. As I watched, the Streblomastix gave a sudden convulsive twist that carried it out of the field of view.

We watched for perhaps twenty minutes until the preparation eventually died, probably through the seepage of poisonous oxygen. Bill described and named one after another of the strange creatures we saw. It was his research project, a.k.a. hobby, to classify and describe the inhabitants of the dark recesses of the termite. Every now and then in the past, he had selected a species with an especially intriguing anatomy for further investigation. Fixed and embedded in resin, the creature would be cut into ultrathin slices. Sections of its anatomy would be viewed in an electron microscope—a procedure for which Bill was justifiably famous. Many of these pictures revealed new microanatomical structures, especially those associated with flagella. But what impressed me most deeply—the lasting memory I have of this visit—was the sudden view it gave me of this hidden world, teeming with life. Why were these strange creatures living in such an unlikely place? Why were they moving? Where was there to go? Who was eating whom, and why?

A few years later Bill retired from the Zoology Department and bequeathed to me a pile of videotapes he had made of the organisms in termite guts. As I watched these tiny animals writhing and crawling through their hidden world, it occurred to me that I was seeing them through a screen of science. I knew (sort of) what was occurring in each waving flagellum. I knew (in broad terms) of which chemicals the beast was made, how it generated energy, and how it sensed its environment. But this information, acquired from books and research papers, gave me no clues to the motive forces and internal states of these living forms. What if these same images were shown to someone who knew nothing of microscopes or modern biology—perhaps from an Amazonian tribe with no knowledge of modern civilization? What would he or she make of these strange wriggling forms? Surely they would seem monsters from a nightmare world, moving with a purpose driven by dark motives. Even a sophisticated Victorian microscopist, meticulously noting the morphology and classification of protozoan species, might speculate (as indeed many did) about the psychic properties of these "infusoria": whether their behavior was in any sense "conscious." But of course we, in this molecule-besotted, fact-filled twenty-first century, know better . . . or do we?

The kind of naïve natural history observation that so fascinated Bill is deeply unfashionable today. You would find it difficult to get a research grant to study the morphology and behavior of protozoa for its own sake. But a century ago it was cutting-edge science. Eager biologists equipped with shiny new microscopes of unprecedented power devoted their careers to observations of the miniature living world. Thanks to them we know that every corner of our Earth is fertile, full of life. In a charming passage in *Manual of the Infusoria*, William

Saville-Kent, formerly assistant to the famous English biologist T. H. Huxley, revealed his enthusiasm:

> On Saturday, October the 10th, 1879, a day of intense fog, the author gathered grass, saturated with dew, from the Regent's Park Gardens, the Regent's Park, and the lawn of the Zoological Gardens, and submitted it to microscopical examination, without the addition of any supplementary liquid medium. In every drop of water examined, squeezed from the grass or obtained by its simple application to the glass slide, animalcules in their most active condition were found to be literally swarming, the material derived from each of the several named localities yielding, notwithstanding their close proximity, a conspicuous diversity of types.

This diversity Saville-Kent then proceeded to enumerate in painstaking detail. No surprise, because we now know that living forms are everywhere—meters down in soil, suspended in the surface waters of the oceans or in their muddy sediments, embedded in Arctic ice, even floating in clouds. Every crack and cranny of our urban environment is a universe where forms visible only under a microscope crawl, swim, compete, and struggle for existence. We are aware of this fact; it conditions our daily hygiene. But what do we really know about the life of these organisms? What do they sense? How do they respond? What is important to them?

These are difficult questions and, like the subject of protozoa behavior itself, extremely unfashionable. Indeed, there seems to be an unwritten convention or law that one should not even raise these issues in a scientific context. Contemporary biologists have an amazing ability to visualize and record what happens in cells. They not only follow single free-living cells but also identify cells moving in the depths of an embryo or in adult tissues. They can pick out specific structures or even single molecules, watch as they move from one location in a cell to another, probe them with microelectrodes or laser tweezers. You can find videos of moving cells on the Internet. But you will be hard put to discover, in all this amazingly rich resource, anyone prepared to ask, as Barbara McClintock did in her Nobel acceptance speech, what knowledge a cell has of itself.

And that, surely, is to be regretted. We have such an abundance of knowledge about living organisms, certainly compared with what the Victorians knew, that we should surely be able to tackle this fundamental question. Like manic pathologists at an autopsy competition, we have littered our workbenches with the dissected viscera of cells. Functional parts (organelles) and molecules of all kinds are set out in display, minutely described and labeled. But where in this museum of parts do we find sensation, volition, or awareness? Which insensate substances come together, and in what sequence, to produce sentient behavior?

Addressing these issues in this book will take us on a voyage that visits most corners of contemporary biology, from protein chemistry to psychology and beyond. But let us start by defining exactly what it is that single cells can do and by tracking the simplest animate wanderings.

Most bacteria are simple rod-shaped cylinders, a few microns long. One micron is a millionth of a meter, a thousandth of a millimeter; a human hair is about eighty microns in diameter. It would take thousands of bacteria to cover a period on this page. They have a tough outer wall and a cytoplasm containing a jumble of protein, DNA, and other molecules. Despite their small size and rudimentary construction, bacteria are capable of independent locomotion, by swimming or gliding over surfaces. In 1854 the German biologist Wolfgang Pfeffer showed that if he introduced a capillary pipette filled with nutrient mixtures such as yeast or meat extract into a solution containing swimming bacteria, the bacteria would collect around the pipette and eventually enter into its tip. Capillaries filled with acid, alkali, or alcohol had the opposite effect, causing the bacteria to swim away. Other investigators at the time observed bacteria responding to light, temperature, or the concentrations of salts. Interest in this simple and accessible system then lapsed for many years. It was not revived until the 1970s, when Julius Adler at the University of Wisconsin in Madison began to systematically analyze food seeking in the common gut bacterium *Escherichia coli*. Today, thanks to the discoveries of a generation of microbiologists, biochemists, geneticists, and biophysicists, we have a detailed knowledge of the molecular machinery of *E. coli* chemotaxis (that is, movement toward chemicals). No other form of animal behavior is understood at anything like this level of detail.

Bacteria swim by means of thin helical flagella, like curly hairs on their surface. Driven by tiny molecular machines—literally motors—embedded in the bacterial membrane, the flagella rotate at speeds of more than one hundred cycles per second. The motors sporadically stop and start, change their direction of rotation, and in this way steer the bacteria according to their surroundings. In the case of *E. coli*, the motors spend most of their time spinning in a counterclockwise direction. When they all turn the same way, the four to six flagella collect into a tight helical bundle, like a pigtail, that drives the cell in one direction through the water. But every now and again, at a frequency that depends on the local environment, one or more of the motors switches to a clockwise direction. This brief reversal breaks up the flagella bundle, and the cell performs a brief chaotic dance called a tumble. Coming out of a tumble, the cell heads off in a new direction. Which way it goes is uncontrolled, random: what is important is not where it goes when it tumbles, but when.

Escherichia coli can detect something like fifty distinct chemicals. The list includes sugars and amino acids that act as attractants (the bacterium swims toward them) and a motley mixture of heavy metals, acids, and toxic substances that are repellents. *E. coli*'s sensitivity is legendary. Even the slightest whiff of the attractant amino acid aspartate (a concentration of less than one part in ten million) is enough to change its swimming. A cell detects a substance that sticks specifically to its surface—the stronger the binding the greater the sensitivity. In the case of aspartate, just a few molecules are enough to turn the cell.

The molecular mechanism of *E. coli* chemotaxis is a superb illustration of cellular information processing. But the most salient point to mention here is that the movements of the bacteria are highly unpredictable, or "noisy." These tiny cells are continually buffeted by water molecules and easily knocked off course, a universal aspect of all small particles suspended in water. So to pursue any direction for more than a second or so, bacteria have to continually reassess their situation. How do they do this? The answer is that they have a sort of *short-term memory* that tells them whether conditions are better at this instant of time than a few seconds ago. By "better" I mean richer in food molecules, more suitable in acidity and salt concentration, closer to an optimum temperature, and so on. If on average conditions have improved, or at least are not any worse, then the bacteria will continue to swim in the same direction. But if conditions are deteriorating, then the bacteria tumble; they swim off in a new direction, selected more or less at random. The repeated execution of this pragmatic routine carries them over long distances and complicated terrains toward favorable locations.

But what do I mean by saying that bacteria have a short-term memory? Doesn't this phrase assign to bacteria a capacity that is really found only in higher organisms? Words like *memory, awareness,* and *information* are easy to use but require careful definition to avoid misunderstanding. I'm using *short-term memory* here in a colloquial, nonspecialist way, referring to how a swimming bacterium carries with it an impression of selected features of its surroundings encountered in the past few seconds. This continually updated record is crucial for chemotaxis, because without it the bacterium would not be able to tell whether it was moving toward or away from a more favorable environment.

And how do I know bacteria have a memory? You can demonstrate it in the following way. Take a population of bacteria in a small drop of water and measure the fraction of cells that are tumbling at any instant. For a typical strain of *E. coli* this fraction will be perhaps 20 percent. Now add a minute quantity of aspartate. The cells will immediately suppress their tumbles, and the fraction of tumbling cells will fall to close to zero. In other words, the cells have experienced an improvement in their environment (a taste of food) and consequently persist in their current direction of swimming. Adding the food substance uniformly to the solution makes every direction equally advantageous.

But now observe what happens to the swimming bacteria. Over the next minute or so, you will see first one then another bacterium start to tumble. Eventually (after a time that depends in part on how much aspartate you added), every one of the bacteria will be swimming and tumbling just as though nothing had changed. Once again, approximately 20 percent will be tumbling at any moment. So if this were your first view of the cells, you would not know that they were now immersed in aspartate. (If you're worried about the fact that the bacteria eat aspartate, then the same experiment can be performed with substances that are not devoured by the bacteria.) But the bacteria are indeed changed by their experience, as can be shown simply by removing the attractant. Immediately, all of the bacteria (or almost all, since these responses are highly variable and no two bacteria are *exactly* the same) begin to tumble, frenetically and without interruption. Evidently they have sensed that conditions have now deteriorated. They have changed their swimming pattern to move to a different location rather as you or I might move away to avoid an unpleasant smell.

So I can state that when the bacteria came in contact with aspartate, they became subtly changed. They acquired an internal trace or record that remained even after the visible effects on swimming caused by the aspartate disappeared. This trace or record corresponds to what is termed adaptation in behavioral experiments, and it represents a sort of knowledge acquired by a cell. Note, however, that it is not the same thing as learning. The bacteria always do the same thing given the same set of environmental stimuli. A biologist would say that their responses are "programmed by their genes," or, more simply, "hard-wired."

The molecules controlling the behavior of the bacterium—mainly proteins—are made according to instructions inscribed in their genetic material, or DNA. After all, bacteria were not taught their behavior; a baby bug does not acquire orienteering skills at school. Nor do they acquire their skills by a process of trial and error, since each individual knows unerringly what to do. They perform the same ritualistic movements characteristic of their species, in response to the same set of stimuli, as did their parents. Their offspring will in turn perform in essentially identical fashion.

In other words, these are automatic reflexes, inherited from generation to generation. And where did they come from in the first place?

Was it because some patriarchal bacterium experienced a blinding flash of insight, an epiphany that it passed to its descendants? Hardly. What must have happened is that certain random changes occurred in the DNA of this cell. Changes in DNA led to slightly different proteins being made, since the genes of an organism specify the structures and functions of all of its proteins. In our Abraham bacterium the new proteins resulted in new functional connections being made. The new protein circuitry caused the cell to behave in a new way. This gave the cell an advantage in a tricky situation so that it survived when its compatriots perished. Its DNA, containing the blueprint of the novel behavior, was replicated and passed onto subsequent generations.

Our memories are stored in the brain. They are represented there by what was once referred to as the engram, or a constellation of connections between nerve cells. Each nerve cell is capable of generating electrical signals, because of special proteins in its membrane (the thin oily skin that encloses every cell). Specific connections, called synapses, allow signals to pass from one nerve cell to another. Sets of nerve cells connected in this manner establish complicated electrical circuits that communicate and process information, analogous to those in a computer or other electronic device. Changes in these nerve circuits constitute a memory. When you learn a telephone number, for example, synapses between certain nerve cells deep in your brain change, becoming stronger or weaker. These changes are, in turn, encoded in protein molecules that change their physical and chemical state or their location in the cell.

The storage of memories by higher animals and the effect on the animals' movements are highly dependent upon the training regime and the internal psychological state of the organism. Complex mechanisms allow the strengths of synapses to change with experience. They are no longer specified solely by the DNA but also influenced by recent events. This is "learning."

The bacterial memory, by contrast, is highly predictable and stereotypical. Whether there is in this single cell anything that could be termed a psychological state is a matter for debate; there is in any case little evidence that this can affect the response to aspartate. But from a superficial, operational sense the bacterial memory and the short-term memory of a higher animal perform similar functions. They also have common elements at a deeper mechanistic level because both entail modifications of proteins associated with a cell membrane. Therefore—if I can do so without invoking a large amount of psychological baggage— I would like to use *memory* as the most simple and most easily understood term.

Bacterial behavior seems primitive when compared to that of protozoa. Organisms such as *Paramecium, Stentor,* and *Amoeba* are also single cells, but they differ in being eukaryotic. The transition between bacteria and their relatives and all other organisms was one of the most significant events in the history of life on Earth. Bacteria are classified as prokaryotic, a word meaning "before the nucleus," because they have a relatively simple internal organization that lacks a nucleus. The cells of plants and animals, by contrast, are eukaryotic, meaning that they contain "true nuclei." They are also larger than bacteria and have a much more complicated interior—subdivided by membranes into different functional compartments, or organelles. The greater size and complexity of eukaryotes equip them for a much larger range of lifestyles than are available to bacteria.

The directed movements of single-cell eukaryotic organisms, or protozoa, were a source of great fascination to microscopists, beginning with the pioneering days of Anton Leeuwenhoek in the seventeenth century. The most productive era was probably in the latter part of the nineteenth century and the early twentieth century. In this period, such biologists as Herbert Jennings, Jacques Loeb, and Samuel Mast in America and Friedrich Schaudinn and Max Hartmann in Germany produced beautifully written and illustrated accounts of single-celled "infusoria" that are classic examples of the art of observation. Armed with nothing more than a microscope and his own keen sight and intelligence, each was able to extract essential features of single-cell behavior and to draw rational conclusions about them. It is difficult for us in an age of instant images and streaming videos to recapture what it meant to be able to peer though a lens and see in illuminated detail a form of life that was otherwise hidden—rather like the "quiet parlor of fishes" Henry Thoreau saw beneath the frozen surface of his pond, but with the added zest of discovery and advancement. To this cohort of scientists the existence of this hidden life was a revelation, to which they alone were privy—one that raised fundamental issues about the origins of movements and sensibility in living cytoplasm.

If one organism can be said to stand for this school of biological investigation, it is the genus *Amoeba*. These single-celled creatures are relatively large (up to one millimeter in some species) and technically easy to observe for hours. Most types of amoeba are predatory: they spend their time in pursuit of other diminutive living forms, a mode of life that evidently calls for an extensive repertoire of movements. But perhaps it was the amorphous form and transparent body of amoeba— the name comes from a Greek word meaning "to change"—that made it a favorite topic for philosophical speculations. How is it possible that a mere transparent blob of watery gel (terms like *protoplasm* and *primordial slime* were also used) is able to crawl around and catch food? A good question, you might say, since we are still not able to give a proper answer.

A single cell of *Amoeba proteus* is a tiny speck of cytoplasm less than half a millimeter in length. Its amorphous, highly changeable form features fingerlike projections—pseudopodia—from a central mass. The cell lives on the bottoms of streams and ponds and typically crawls over the surfaces of pebbles and vegetation by extending and retracting its pseudopodia. When it encounters obstacles in its path, regions in contact with the obstacle become quiescent while those farther away move more actively. Consequently the cell flows in a new direction. Other modes of movement are also seen. For example, a hungry amoeba often adopts a more rapid form of progression termed loping, alternately extending and contracting its body like an inchworm. Another state I like to think of as skydiving occurs when an amoeba is freely suspended in water. The cell now adopts a starlike form with long pseudopodia extending in all directions. It seems to be searching for a suitable surface. Indeed, no sooner does one pseudopodium make an effective contact than currents of cytoplasm flow in that direction. Other pseudopodia

retract; the cell settles onto the surface and resumes its normal mode of
creeping.

Amoeba proteus eats minute plants and animals it finds in its path. It
has a strong preference for certain foods and changes its mode of feeding according to what is available. Herbert Jennings, a naturalist working at the Zoological Laboratory of the University of Michigan in the
late nineteenth century, observed spherical cysts, or spores, of *Euglena*
(a ciliated protozoan) being used as food. The amoeba in his cultures attempted to engulf the cysts by extending lateral pseudopodia and a
thin sheet of cytoplasm. If the attempt was successful, then the cyst was
enclosed in a cup-shaped depression, entered the cell in a parody of swallowing, and was eventually digested. The cysts, however, were smooth
and rolled easily when touched. Jennings recorded an amoeba that pursued a rolling cyst in a circular path for fifteen minutes, making repeated
attempts at capture, until the cyst was eventually lost. "This behavior,"
Jennings remarked ". . . makes a striking impression on the observer
who sees it for the first time. The Amoeba conducts itself in its efforts to
obtain food in much the same way as animals far higher in the scale."

Rather surprisingly, *Amoeba proteus* is also able to capture actively
swimming organisms such as the ciliated *Paramecium* despite its lumber-
ing gait. It seems to be able to detect swimming paramecia at some distance and may pursue them over long distances. Once contact is made,
the amoeba initiates a slow encircling action and attempts to trap the
struggling prey. If encounters with swimming prey are sufficiently frequent, the amoeba anticipates a capture by forming a cup-shaped depression on its surface like a baseball mitt.

The German naturalist Johann Rhumbler described a more elaborate form of feeding. The smaller species *Amoeba verrucosa* lacks
prominent pseudopodia. These organisms are fond of filamentous
green algal cells, which they manage to devour despite the prey's
being many times their own length. In a typical feeding sequence, an
amoeba first settles upon the middle of the filament and wraps around
it before spreading in either direction along its length. One end of the
amoeba bends over and forms a loop in the algal filament. Then the
amoeba stretches out on the filament and bends it anew. This process

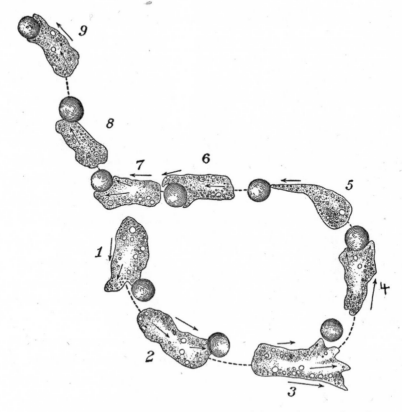

FIGURE 1.1. An amoeba attempts to capture and devour the cyst of another
protozoan—a potential source of food. From Jennings 1906.

is repeated until the alga forms a compact coil within the body of the
amoeba and is digested.

In 1908 David Gibbs of Clark University published a research article entitled "The Daily Life of Amoeba proteus" in *American Journal
of Psychology*. A strange place to find a study of protozoa, you might
think, but indicative of the way these simple creatures were viewed at
the time. Gibbs's objective was to discover whether amoebae have distinct phases of activity comparable to the rhythms seen in higher animals.
He recorded the activities of individual cells for periods of up to four
days and nights and saw no evidence that they slept at regular times (it's
not clear how much sleep he had himself). But Gibbs did find that the
level and nature of movements varied from hour to hour. Amoebae

would persist in such activities as "feeding on algae," "traveling with many pseudopodia," "loping," "resting," "dividing," for several hours on end. It was as though the organisms adopted distinct internal states, analogous to being hungry, inquisitive, or tired.

Remarkable though they are, amoebae are not at the pinnacle of single-cell complexity. This position is occupied by ciliates and flagellates. Characterized by the presence over much of their surface of long waving hairlike cilia or flagella, these organisms include the strange and exotic inhabitants of a termite gut described at the beginning of this chapter. Many ciliates, such as the slipper-shaped *Paramecium,* are excellent swimmers and dart around vigorously sweeping particles of food (bacteria in this case) into their mouths. Paramecia avoid obstacles by a stylized square dance in which they reverse direction, turn through a fixed angle, and again swim forward—repeating this process as necessary. (I will describe a similar mode of obstacle avoidance in an electronic robot turtle in the next chapter.) They also have sex by fastening onto other paramecia and exchanging genetic material. A different ciliate, *Didinium,* swims around like a high-speed syringe, spearing other cells with its sharp proboscis and sucking out their contents. Yet other forms employ groups of cilia as fins or even as legs to crawl over surfaces. Only a small fraction of these creatures have been studied in detail; much surely remains to be discovered.

Unexpectedly, the most intricate behaviors so far recorded in unicellular animals have been found in organisms attached to a surface rather than swimming free. The trumpet-shaped *Stentor* spends most of its time attached to water plants or debris by an irregularly shaped foot embedded in a transparent tube of mucus. Its slender body and flattened, discoid mouth cavity are lined with fine cilia. Beating of these cilia sets up continual currents of fluid that carry particles of suitable size into the mouth cavity. The animal then decides whether they are edible or not. Particles of food are retained and ingested, whereas indigestible particles are ejected. The organism can also turn or sway on the end of its long stalk, thereby pointing its mouth cavity in selected directions.

The lynx-eyed Herbert Jennings gave a detailed account of the response of *Stentor roeselii* to experimental stimulation. A fine current of

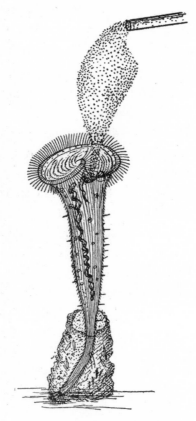

FIGURE 1.2. A cloud of red dye particles
is introduced into the mouth of
Stentor, a sessile protozoan, from
a pipette. From Jennings 1906.

water squirted at the open disk caused the cell to contract in a flash into
its tube. After about half a minute the animal extended again and the
cilia resumed their activity. A second application of the jet of water
identical to the first, however, was then ignored, and the animal continued with its feeding activities. Jennings observed a similar sequence of a
"surprise" response followed by rapid acclimatization with a whole variety of stimuli, such as a small jarring of the animal's container. Once
again the organism displayed a simple form of memory.

Jennings found, however, that a potentially injurious stimulus, such
as a cloud of particles of the red dye carmine introduced to the feeding

disk, evoked a richer set of actions. The stentor at first makes no response, taking the obnoxious particles of carmine into the pouch and
into the mouth. After a short time, though, it turns to one side by
bending its long stalk (always in the same direction) as if to move its
mouth out of the path of the cloud of noxious particles. This may be
repeated: the animal twists on its stalk two or three times about its long
axis. If the repeated turning to one side does not relieve the irritation, a
second strategy is tried. Beating of the ciliary hairs on the body is suddenly reversed in direction, thereby causing particles in the disk and in
the pouch to be ejected. The reversal lasts but an instant before the
usual current is resumed. If the irritant particles continue to arrive, the
reversal is repeated two or three times in rapid succession.

If stentor still does not get rid of the stimulation in either of the
ways just described, it contracts into its tube. In this way, of course, it
escapes the offending cloud, but at the expense of losing all opportunity to obtain food. The animal will now remain in the tube about half a
minute before again cautiously extending. When the body has reached
about two-thirds its original length, the cilia around the disk-shaped
mouth begin to unfold. They recommence their beating, allowing the
animal to recommence feeding.

What happens now? Suppose that the water currents again carry noxious carmine grains into the feeding disk. The stimulus and all the external conditions are the same, so will the stentor go through the same
routine—bend to one side, reverse its cilia? No. The animal has become
changed by its experience and tries a new strategy. This time, no sooner
do the carmine particles reach its disk than the animal contracts fully into
its tube. The sequence may be repeated many times, the animal cautiously
extending after a while and, if it still finds particles, contracting again.
Each time, it stays a little longer in the tube and seems to be more reluctant
to emerge. Finally, it contracts violently and repeatedly while still enclosed
in its tube, so violently that the foot attachment breaks and the animal is
set free. Stentor leaves its tube and swims away, avoiding if necessary
the cloud of carmine and other obstacles, and sets sail for a new home.

In due course, the minute creature will settle onto the bottom of
its pond or container. It then behaves in a peculiar way. The partially

unfolded disk (the part previously used for feeding) is now used as a means of locomotion. Stentor now creeps rapidly over the terrain, following all the irregularities of the surface rapidly and seeming to explore it thoroughly. This may last for some time, the animal often quitting one site and swimming on to another, exploring heaps of debris and surfaces of solids in its path.

Perhaps ten or twenty minutes later, one location is selected for the formation of a new tube. As the stentor moves around, viscous mucus is secreted over the surface of its body. Particles of debris sticking to the mucus are trailed behind the animal. After a while, the animal stops and begins to move repeatedly backward and forward. This motion compacts secreted mucus into a tube—making a sheath in which the stentor can live. Next the foot presses against the debris at the bottom of the tube, where it adheres by means of fine pseudopodia sent out from the body. Finally, the stentor extends to full length and adopts its familiar condition, with the lower half of its body enclosed by a transparent tube of mucus. It has now successfully moved away from the place where it was subjected to irritating, noxious particles and has established itself in a new home.

This entire pattern of actions is performed by a fluid-filled sac of membrane less than a millimeter in length!

At the beginning of the twentieth century, then, biologists knew that single-celled organisms are capable of complicated sequences of actions in response to a wide variety of stimuli. All free-living cells, including bacteria, amoebae, and ciliates, can detect chemicals in their surrounding media. They achieve this sense of taste and smell, as we do with our sense of smell, because molecules in the outside world stick specifically to their surfaces. Signals generated by proteins in the membrane then tell the cell about possible sources of food and potentially damaging environments. Amoebae crawl over surfaces and steer past obstacles they cannot surmount. Species such as *Vorticella* (a close relative of stentor) collapse rapidly in response to minute tremors in the water. In the world of the very small, the beating of a cilium can send vibrations over distances of hundreds of cell diameters. Cells could use the magnitude and rhythms of these waterborne vibrations to gain a sense of what is in their neighborhood, rather like a primitive form of hearing. Almost all single-celled organisms respond to light. Ciliates such as paramecia move to weak light but are repelled by strong light. The single-celled alga *Chlamydomonas* has a so-called eyespot containing pigment molecules that enable it to detect the direction of incoming light. There is even a ciliate called *Erythropsidium* that spends its life attached to the bottom of a pond, watching the world through a large eye equipped with a lens! Strange indeed, since in most eyes a lens serves to focus images onto a retina, an outpocketing of the brain. There is no retina here, no brain.

In other words, single cells are aware of their surroundings. They detect chemical flavors, mechanical vibrations, visual stimuli, electric fields, and gravity. They respond by moving or by changing their shape or internal state, selectively, in a discerning manner. An amoeba can immediately tell a *Euglena* cyst from a grain of sand of the same size and will devour the former while rejecting the latter. Swimming paramecia continually encounter different situations that they either accept by going ahead or reject by reversing. Almost any animate wanderer must make decisions. When confronted with multiple stimuli of a possibly conflicting nature, cells have to evaluate their options and assign priorities. They must actively choose one out of many possible responses.

Most microorganisms display what in higher animals is termed attention. A stentor or amoeba dislodged from a surface actively seeks a new site of attachment. While engaged in this search it ignores other stimuli such as changes in temperature or chemical signals that produce an immediate reaction in a free-living individual. Indeed, it is probably necessary for something like attention to exist, since it is usually impossible for a cell to react simultaneously to two or more kinds of stimuli. I use *attention* here in a colloquial sense, as I did *memory* above, to avoid the psychological ramifications. This is not the same as human attention.

The repeated or prolonged application of the same stimulus to a single-celled organism frequently evokes a response that changes with time. Recall that this happens when *E. coli* bacteria encounter an attractive substance. Their immediate response is to suppress their tumbles, but if the stimulus remains constant over a minute or so, the tumbling returns. Similarly, stentor initially responds by violently avoiding a light jet of water but will soon ignore the stimulus if it is continued. Changes of this kind are described as adaptation or accommodation but also constitute a kind of memory.

If we had observed these patterns of movement in a higher animal, then we would have attributed them to a subjective state of the organism, employing such words as *hunger, pain,* or *fear.* That they occur in minute scraps of living matter visible only under a microscope changes our usage. Indeed, it seems an unwritten rule that any description of lower forms of life should studiously avoid any mention of subjective states. Apparently one should describe them as though the objects in question were inanimate. But is this justified? Is it not natural to ascribe to these creatures internal states and feelings?

Victorian microscopists argued at length about these issues and expressed most of the possible viewpoints. To a deeply religious observer such as William Saville-Kent the diversity of forms and movements under the microscope was a tribute to the Creator. He had little reason to inquire further into origins and motives, since to him they were divinely inspired. Alfred Binet, the French psychologist who introduced the term *intelligence quotient* (IQ) to the vocabulary, was a vitalist, believing that living organisms possess unique but mysterious properties. In his 1888 book *The Psychic Life of Microorganisms* he assigned sensations and motivations to even the simplest single-celled organisms. He declared that psychological life began with protoplasm and arose from "an aggregate of properties which properly pertain to living matter and which are never found in inanimate substances." Such views were slowly displaced by more mechanistic views of physiologists such as the American Jacques Loeb. In his 1918 book Loeb declared: "Since we know nothing of the sentiments and sensations of lower animals, and are still less able to measure them, there is at present no place for them in science." Most present-day scientists probably hold the same opinion.

The aforementioned Herbert Jennings took the middle ground—to my mind the most interesting position. In his 1906 book *The Behavior of the Lower Organisms,* Jennings admitted that subjective experiences are, almost by definition, inaccessible to objective measurement. But this did not stop him from asking questions. Does an amoeba perceive all classes of stimuli that we ourselves perceive? [Yes.] Are the responses of amoebae to such stimuli of a similar kind to those of conscious organisms? [They are.] Would it not assist an amoeba, a beast of prey, if it indeed were controlled by impulses and emotional drives analogous to those of higher organisms? [He suggests that it would.] Indeed, Jennings wryly adds, if an amoeba were as large as a whale, we humans would be well advised to anticipate its movements on the basis of internal states, such as appetite!

What exactly did Jennings mean by such words as *perceptions, impulses,* and *emotions* when applied to single cells? Was he suggesting that these scraps of living matter might have internal, subjective states that are in any way comparable to those experienced by humans? And tied up with that adjective *subjective* is the tricky word *consciousness.* If there is a subjective experience of pleasure or pain, does there not have to be an internal observer? Implicit in the questions raised by Jennings is the notion that if by magic I could inhabit the body of a microorganism, then I would experience recognizable feelings and emotions. But does this make sense? Is it conceivable that these simple organisms could possess an awareness of their surroundings that is in any way comparable to our own?

These are not new questions. Related issues have exercised philosophers over the centuries since the ancient Greeks. Descartes, Spinoza, and Leibniz all pondered at length on the nature of animal emotions and sentience, on the similarities and differences these show to the sensations of humans. The views of René Descartes (1596–1650) were especially influential in this regard. His theory of the fundamental separation of mind and body remains a foundation of Western intellectual thought. In *Treatise on Man,* Descartes declared that both animals and humans are wonderful machines. Developing this argument, he imagined a hypothetical humanoid machine that operated by clockwork or hydraulics, like the clockwork automata that were then being built in chateau gardens and clocktowers. Similarly anatomists of the time, such as Thomas Willis, frequently adopted analogies of a mechanical kind for internal organs and bodily structures, mentioning cords, ropes, pumps, and valves. So parallels with animated statues were irresistible. "From this aspect," Descartes wrote, "the body is regarded as a machine which, having been made by the hands of God, is incomparably better arranged, and possessing in itself movements which are much more admirable, than any of those which can be invented by man."

Although the bodies of both animals and humans can be seen as machines, he said, only humans have a mind—the capacity to think and speak. This had to be so, Descartes argued, because if horses, honeybees, and swallows were able to think then they would possess immortal souls like humans, and that was impossible. Curiously, from our standpoint, he did not deny that animals possessed other natural impulses such as anger, fear, and hunger. He did not question, apparently, the universal nature of these biological drives or the common features by which they are experienced. It was by their "rational soul" that humans were set apart, although exactly what he meant by *soul* is a matter of debate. The power of the church in the mid-seventeenth century was oppressive; the arrest of Galileo by the Catholic Inquisition was a warning to all freethinkers of the risks of heresy. But it is ironic that Descartes used an automatic humanoid machine to frame his philosophical arguments. Modern-day philosophers continue to wrestle with similar issues: Can machines think? Will robots ever be conscious?

To René Descartes the origins of animal behavior were a mystery, part of the vital force that distinguished living things from inanimate matter. Little point his asking about subjective experiences when so much else was in darkness. But by 1906, when Herbert Jennings's book was published, he would have understood many of the basic mechanisms of behavior. There was already widespread recognition of the chemical basis of biological organisms. The Swedish chemist Jacob Berzelius (1779–1848) had seen "animal chemistry" as a worthy goal and declared his aspiration "to find a clue to the chemistry of the living body through the chemical knowledge of our laboratories." Substances previously thought to be unique to living organisms had been manufactured in the laboratory. Investigations were under way of the seemingly mysterious abilities of biological extracts to cause chemical change, such as in the production of alcohol from sugar. An appreciation of the crucial substances Berzelius named proteins was starting to emerge.

Jennings would also have known of the cellular basis of living organisms, which was widely accepted by the end of the nineteenth century. His contemporaries included Edmund Wilson, professor of zoology at Columbia University and author of the seminal *The Cell in Development and Inheritance,* and the Spanish microscopist Santiago Ramón y Cajal, who, more than anyone, demonstrated that the nervous system of higher animals is made of cells. The electrical nature of nerve activity, presaged by the findings of Luigi Galvani in the eighteenth century, was also widely accepted. Using fine electrodes and sensitive recording apparatus, investigators could demonstrate that nerve cells are connected electrically: they detected electrical signals traveling down axons, passing to other cells across junctions called synapses.

As these discoveries unfolded, they led to profound changes in our view of animal movements. Layer by layer, the underlying processes were revealed. Motor reflexes and the action of voluntary muscles; sensory processes such as vision, hearing, touch, and balance; the hormonal and neuronal basis of emotions including fear and anger; mental processes such as pleasure, sexual arousal, memory, and learning—all were laid bare. In broad terms, these could be explained as arising from chains of nerve cells linked by synaptic connections. The machinelike nature of animal behavior became part of popular psychology.

Consider the following example, cited in Dean Woolridge's influential 1963 book *The Machinery of the Brain.* The solitary wasp *Sphex,* being pregnant, digs a burrow to accommodate her eggs. She then seeks and captures a cricket, cleverly stinging it in such a way that it is paralyzed but not killed. Dragging the cricket into the burrow, Sphex lays her eggs by its side, carefully seals the burrow, and flies away. When in due course the eggs hatch, the grubs feed off the cricket, which has been kept hermetically fresh in the interval, and eventually they emerge. To a human observer this sequence appears deliberate and sagacious; the mother wasp seems to have a plan of action—an understanding of the world and her part in it.

Closer examination reveals this to be an illusion. The wasp's normal routine on returning with the paralyzed cricket is to leave it on the threshold of the burrow and then go inside—her hunting trip may have taken some time and she needs to check that it is empty of other inhabitants. All being well, she then returns to the entrance and drags her paralyzed booty into the den. But if a human observer intervenes at this juncture and deliberately shifts the cricket a short distance while the wasp is in the burrow, then the routine goes awry. Instead of simply retrieving the cricket and dragging it forthwith into the burrow, as would appear sensible to a human observer, the wasp replays the previous sequence of actions. She drags the prey to the burrow entrance, then enters the burrow again, seemingly ignorant of the fact that she has already checked its security. If the cricket is then moved a few inches, the wasp will once again pull it back to burrow entrance and reenter for another check. The wasp never learns . . . never realizes that she has just checked the burrow and so pulls the cricket straight in. On one occasion, researchers observed the operation repeated forty times, always with the same result.

There are many other examples. A worker bee that has found food will execute a characteristic wagging dance when it returns to the hive, signaling to other bees the direction, distance, amount, and quality of the distant source. But she can be tricked into performing the same dance in isolation and in the complete absence of other bees, simply by tickling her antennae. This stimulus is sufficient to trigger the dance, even though it is now meaningless. Similarly, a silkworm interrupted halfway through the process of spinning her cocoon by the removal of her handiwork resumes the process as though nothing had happened. She dutifully finishes an incomplete cocoon that is nonfunctional.

I'm not saying here that invertebrate animals are incapable of learning. The Nobel Prize–winning work of the New York scientist Eric Kandel on modifiable synapses, a crucial element in learning, was performed entirely on the sea slug *Aplysia*, a primitive organism with no backbone. But it is undeniable that a major portion of their behavior is hard-wired in the same sense as bacteria. Genetic instructions inscribed in the sea slug's DNA direct the connections made between nerve and muscle cells. Their responses are therefore repetitive and predictable.

Deeper analysis reinforces the impression of blind circuitry. Invertebrate animals have far fewer nerve cells than their vertebrate counterparts, and these nerve cells are often relatively large. It is often possible (for example, in nematode worms, fruitflies, and other species) to identify the same nerve and muscle cells from animal to animal. Cells can be given names, relocated time after time under the microscope. They can be penetrated by microelectrodes and their characteristic electrical responses to specific stimuli recorded. Experimenters can dissect out sets of interconnected nerve and muscle cells and examine their reflexes independently of the rest of the organism. In this way, the circuitry of simple activity patterns can be charted like an electrical circuit, with all of the essential components present. An isolated leg of a cockroach removed from the rest of the insect but still attached to its nerves and a portion of the central nerve cord flinches whenever stimulated by a heated wire. It is hard to convince oneself that this reflex arises from pain or discomfort. A fragment of a sea urchin carapace carrying some of its multiple eyestalks generates a signal in response to a shadow, but it is difficult to liken this to the human experience of vision. It seems even less probable that this fragment of tissue has assessed the situation and decided to take evasive action.

Single cells, therefore, emphatically do not have feelings or humanlike consciousness. They are just too small and simple. They not only lack a brain with a cortex (which is surely required before any subjective experience that we would recognize is possible), but they have no nerve cells at all! A solitary amoeba is hardly larger or more complex in molecular terms than any one of the thousands of nerve or muscle cells

in a worm or cockroach. If we can reduce the responses of an invertebrate animal to a stereotypical circuit of connections, blindly assembled
during development like the subroutine of a computer program, then
we can surely do the same for an amoeba.

And yet the questions raised by Jennings continue to nag. Yes, we reject
the possibility of consciousness in these simple organisms. Yes, we agree
to call them machines made of biological materials. But does this mean
they have an equivalent status to synthetic robots made of silicon and
metal? Is the similarity of the behavior of paramecia or amoebae to that
of mice and cats just an accident?

For all their small size and (apparent) simplicity, single-celled organisms are nevertheless complete functional systems. I would argue
that any animal large or small that pursues a freely moving existence
must have some minimal level of internal organization in order to survive. It has to detect and recognize salient features of its environment,
move to a suitable niche, detect and hunt down prey, sense and avoid
predators, find a mate. A flood of information enters such a cell every
second of its existence through its membrane. This must be assimilated, sorted, codified. The cell has to choose one integrated, coherent
action that ensures its survival.

Where did all this come from? One possibility is that the requisite
sensory reflexes arose independently on more than one occasion during evolution. Perhaps in single cells one evolutionary path led to reflexes in the form of cascades of biochemical reactions, whereas in
multicellular animals an entirely different historical path led to networks of nerve and muscle cells. Perhaps these were independent
events with no direct relationship between the two. Or might it be—
bearing in mind the seamless web that links all living forms—that
these two strategies are fundamentally related? Is it possible that an
amoeba possesses, in an extremely reduced and primitive form, some
of the mechanisms that mediate the sense of environment and self in
humans? What knowledge does a cell have of itself? Is it completely
dark in there, a black space full of blind molecular machinations? Or

might that watery slurry contain an ember of emotion, a prototype of sentience?

In revisiting these issues, we have many advantages over Herbert Jennings. We now have a much clearer idea of the neural processes that underlie simple forms of animal behavior and even (despite debates and controversy) human consciousness. So it could be that, by working backward from the neuronal circuitry of human brains, we may be able to identify common elements in the biochemical networks of single cells. At the other end of the scale (what would have been even more of a revelation to Jennings) is our present understanding of the molecular basis of life. We now have a rich knowledge of cells and the cornucopia of molecules they contain. We can follow individual reactions, step by step, as a cell detects and responds to its environment, locates and captures prey. Will this not help? Surely by dissecting the molecular base of sensory detection, discrimination, decision, and hunger in living cells, we can learn more about their self-knowledge.

But there is one other source of new information that will guide us. In the twentieth century humans created a new world of computers and electronic robots. Some of these silicon-based machines act and behave in a manner that closely resembles that of living organisms. Researchers today routinely use computers to simulate the activities of parts of the brain and, increasingly, chains of reactions inside cells. We can build robots to replicate the locomotion of salamanders and the communal activities of termites. It seems plausible that this large and active field of endeavor, undreamt of a hundred years ago, might provide essential clues. Perhaps computer-based artificial intelligence can help us understand the world of real organisms. Certainly the general notion of computability will help as we unpack the molecular basis of why cells act as they do.

Simulated Life

A strange form of life discovered tomorrow on a distant planet would transform biology. All of our notions of the origins of life are based on a single example, so it is impossible to tell which features are inevitable and which are historical accidents. Can you have a life form without DNA and proteins? Is it necessary to have large molecules made as polymers (chains) of smaller subunits to encode information and perform chemical tasks? Or might there be radically different solutions to the problems of growth and reproduction, perhaps employing a totally different chemistry . . . or no chemistry at all? This is why an extraterrestrial organism would be such a revelation. By providing a counterexample it would shed light, as though in a hand mirror, on the natural world of Earth.

There are no Martian bugs yet, but we do have thinking machines. Computers, robots, and similar fruits of the human mind, which some visionaries believe will eventually take over the world, have much to teach us about living systems. There is, to begin with, the matter of computation. In a circuit board or microchip, sets of logic elements linked in precise networks perform defined logical processes, repeating the same simple steps over and over again. The same thing happens in a living cell, except that the elements now are molecules instead of transistors and the blind iterations they perform are chemical rather than electrical.

At a higher level, there is the much commented-upon similarity of computers and brains. Both are made from multiple components linked into networks of connections. Both carry digital signals in the form of electrical pulses, process information, store memories. Think of computer toys and games. Those animated pixelated figures on your screen, robots scrambling over your living room carpet, seem almost alive. They are indeed a new form of life and one that has been inordinately successful in numerical terms: there are far more of them than there are elephants or pandas, regrettably. So it makes sense to look more closely at how they are made, why they behave as they do, and why we are so easily duped into thinking they are real.

In 1979 a young Japanese programmer, Toru Iwatani, was assigned by his company, Namco, to create a video game. Toru wanted get away from alien spaceships, guns, and explosions and to design something that women could enjoy as much as men. Inspiration came to him at lunchtime as he was eating a pizza in the company canteen. Having consumed one slice, he thought that the remaining pizza suggested a simple shape (easy to represent in those days of limited pixels) and a story line based on eating. The original name of his invention was PuckMan, derived from a Japanese slang expression meaning to devour. A few years later Midway licensed the game for American distribution, changing the name from PuckMan to PacMan, reputedly to discourage profanity.

In the classic format of the game you use a joystick to move a PacMan around a maze. As PacMan travels he encounters lines of buttons that he scoops up effortlessly in his mouth. Every button adds to your score, and if your PacMan manages to eat them all, you advance to the next higher level. A primordial boy-meets-girl story line unfolds as each level is reached: "They Meet" . . . "The Chase" . . . "The Kiss" . . . "Junior." But your PacMan's passage through the maze is not simple: he is continually challenged four ghosts: Inky, Blinky, Pinky, and Clyde. These chase your PacMan, and if they catch him, you lose a life. When all lives are lost, the game ends. But you can reverse the chase temporarily by feeding your PacMan power food: guide him to one of the flashing beacons in the maze, and when he devours it he acquires, albeit transiently, the power to eat the ghosts.

A metaphor for real life, the game represents not only conflict and survival but also wealth creation. Each button you eat adds to your score, and there are additional treats. Occasionally a pear, a strawberry, a banana, or a pretzel appears and stomps in determined, audible fashion through the maze. If you manage to catch and eat one of these treats without being trapped by a ghost, you add enormously to your score—like receiving an inheritance or making a killing on the stock market. This original simple arcade version, built from flashing pixels on a darkened maze, was in later years superseded by a constellation of slick successors: shiny three-dimensional spheres with cruel mouths, ghosts that sing, pots of gold replacing the strawberries. Successive programmers added a thousand special graphical devices and different scenes. And yet the essence of the game remains.

What is its secret? What makes this and many similar games so addictive for millions of players worldwide? Objectively speaking, it is nothing but a computer screen full of flashing pixels. The obvious answer is that the pixels make shapes that resemble living creatures, albeit in a distorted manner. They appear to move with motivation and purpose. Constrained within the walls of the maze, the ghosts make continual decisions to turn left or right, but always with malevolent intent. Their goal in life is clear: whenever the opportunity arises, they will move in the direction of PacMan, like hungry amoebae closing in on a paramecium. And why am I captivated by this obvious charade? Perhaps because it connects with something in my brain, a primitive association I carry between movement and life. The directed, responsive motility of the ghosts speaks to me of sentience. I have an innate sense that any body that moves, seeks, hesitates, follows, or runs must have an inner life. So strong is this ancient instinct that I am able, briefly and for the purposes of entertainment, to suspend disbelief. This is why, surely, I am able to assign to a system of flashing pixels on a screen the quality of life. It's a long way from the forest or savanna where these reflexes evolved, but the imprint is still there.

There is more to the story. Watch a three-year-old playing with a model dinosaur or a doll and you may see and hear that the child has identified with the toy. "Here I come. Ha-ha . . . I'm going to eat you!

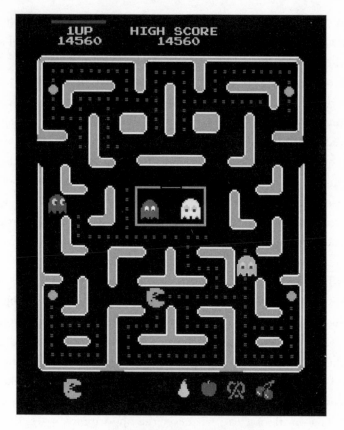

FIGURE 2.1. A game of PacMan in progress.

Don't I look pretty?" Similarly, PacMan, who flees down the maze chased by ghosts, fortified by a steady diet of buttons and an occasional infusion of adrenaline from a strawberry, is none other than your own self in proxy form. As he speeds down the lanes, turning this way and that, PacMan is responding to your mental signals. He is an extension of your body, a continuation of your hand on the joystick, your representative in virtual space. Sophisticated modern programs have realistic representations of realistic human forms that walk and talk on the screen. But the underlying mental equation is the same. Not only does PacMan display the hesitant directed motility associated with sentient life forms, he does so in accord with your mental state.

What has this game to do with living cells? The point is that, to an observer sitting in front of a screen, the movements of PacMan and the ghosts are not so different from those of a predatory amoeba. Each sends unspoken messages to the human brain about the independence of the images. You instinctively know who is the predator and who the prey, and unconsciously assign motivations and even feelings to the images on your retina. But because PacMan, unlike the amoeba, is a human construct, you can know explicitly what instructions it follows. PacMan follows rules written in lines of computer code that specify what he does in any conceivable situation. Somewhere in the watery contents of a paramecium or an amoeba must be something that works in a similar fashion—not an exact replica of the circuitry, not built of the same materials or working in the same way, but its logical equivalent.

The original PacMan game was written in machine code—a highly condensed format necessary because of the severe limitations of computer memory at the time. But we know from other games that just a few hundred lines of code are enough to generate lifelike activity at least at a rudimentary level. In fact, one of earliest and most transparent examples was a physical implementation that has come to be known as Grey Walter's Tortoise.

William Grey Walter was a Cambridge neurophysiologist who, in the early 1940s, found himself the director of a small research institute in Bristol. His main research subject was electroencephalography, the recording of electrical waves from a patient's scalp. This is very much a "view from the outside"—an attempt to deduce events occurring inside the brain from their external influences. Perhaps frustrated by the passive nature of his research tool, Walter began to build, sometime around 1943 and in collaboration with his wife Vivian, mechanical devices capable of independent mobility. They were very simple robots built on a shoestring budget appropriate to wartime England. The Walters used small tubes from a radio, wheels and gears from old clocks and gas meters, and a hearing aid battery. The robots were about the size of a shoebox, each with a dome-shaped plastic carapace mounted on wheels and a single eye on a short protruding neck. They looked somewhat like tur-tles, and Grey Walter dubbed them tortoises, referring to the passage in *Alice in Wonderland* where the Mock Turtle says: "We called him Tor-toise because he taught us." Despite their simple design, the tortoises displayed homing instincts and engagingly hesitant patterns of move-ment.

They built several versions to a similar overall design. The tortoises moved on three wheels, the one in front providing both the steering and the drive. Two independently operating motors were contained be-neath the carapace, one turning the steering wheel and the other driving the robot forward. Sensory input to the tortoise was provided by the light-sensitive eye mounted on its head and by a bump detector mounted between the shell and the chassis. In the absence of an adequate light stimulus—in the darkened arena of its enclosure—the tortoise explored its surroundings. The continually revolving steering motor gave the creature a cyclical crawling gait, while the photoelectric eye looked every direction in turn. This process of scanning and its synchroniza-tion with the steering mechanism enabled the tortoise to remain alert to the possibility of light and to avoid many obstacles. If the tortoise did blunder into an obstacle, its gait became a succession of butts, with-drawals, and sidesteps until the interference was either pushed aside or circumvented. These oscillations persisted for several seconds after the obstacle had been left behind so that the tortoise was sure to move well clear.

When a tortoise saw light—that is, when its photocell generated a signal—the steering mechanism stopped its rotation and the tortoise moved on a straight line toward the light source. Grey Walter compared this to the simple tropism of a moth attracted to a light. (After reading Chapter 1 you might liken it to a bacterium attracted to a distant source of nutrient.) But unlike a moth or a bacterium, the tortoise would never reach the source of attraction. For when the brilliance of the distant light source exceeded a certain value—comparable to that of a flashlight about four inches away—its steering mechanism was activated again, this time at double speed. The creature abruptly sheered away as though search-ing for a more favorable location. Responding to a single light source, the tortoise would circle around, first advancing then withdrawing. If another light were situated farther away, the robot visited first one and then the other, weaving a path between the two.

Grey Walter designed an enclosure for the tortoise that also contained a small kennel. A twenty-watt lamp was attached to the kennel entrance, and inside was a battery charger. When the tortoise's batteries were well charged, it was attracted to the kennel from afar, but at the threshold the intensity of light would be sufficient to activate the repellent response. If the tortoise's batteries ran down, however, its sensitivity to light decreased and its pattern of movements changed. Now if the creature found itself at the entrance to its kennel, it would be attracted to enter. Once inside it could make contact with the charger and replenish its battery. As current flowed into the tortoise, all movement was suspended until the battery voltage had risen to its maximum. Then the response circuits were activated again. The little creature now found the light repellent and pirouetted away for further adventures.

Grey Walter added a small light to the front of the tortoise, connected to the steering motor circuit. His original intention was to display when the steering motor was operating, but then he discovered that it added a curious dance. A tortoise with a light on its head encountering a mirror would flicker and jig as though recognizing its own image. The machine was first attracted to the reflection of its own light, but then, the moment its steering motor stopped, it was sent away by the extinction of its light. Two tortoises meeting face to face were affected in an even stranger manner. Each was attracted by the other's light, but the very act of detection caused it to extinguish its own source of attraction. So the two systems became locked in reciprocal oscillations, leading finally to stately retreat. When the encounter was from the side or from behind, each would regard the other as an obstacle. Tortoises competing for the occupancy of a battery-charging kennel exhibited an even more intricate pas de deux.

Grey and Vivian Walter described other developments of their little creatures. In fact, they even created a second "species" of tortoise. The first version was dubbed *Machina speculatrix* because of its propensity to wander and explore. Its movements were complicated

FIGURE 2.2. Robot tortoises searching for light and avoiding obstacles. From Walter 1950. Art © Estate of Bernarda Bryson Shahn/Licensed by VAGA, New York, NY.

and unpredictable but showed no evidence of learning by experience. In an attempt to rectify this omission, the Walters developed a second type of robot based on the first and called it *Machina docilis* (from the Latin for *teachable*). This had the additional sensory ability of detecting the sound of a whistle. It also had a slightly more elaborate internal circuit that allowed input from the whistle sound to be mixed with the signals from the photocell "eye" or from the bump detector. *M. docilis* could indeed be trained to associate the sound of a whistle with the presence of an external light. The whistle was blown just before the light was seen and, after ten or twenty repetitions, the metal animal would come to the whistle as though it were a light. In a second training scheme the whistle was blown just as the model bumped into an obstacle. After a period of training, the warning whistle by itself could trigger withdrawal and avoidance.

Grey Walter had a liberal, indeed naïve, interpretation of his robots' activities. He drew parallels between his work and the classical studies of conditioned reflexes by the Russian physiologist Ivan Petrovich Pavlov. In these well-known experiments, Pavlov trained dogs to associate a simple reflex, such as a flow of saliva or the withdrawal of a leg, with a conditioning stimulus such as the shining of a light or the ringing of a bell. If a bell were repeatedly sounded before food was presented to a hungry dog, then the dog would eventually salivate (the reflex) on hearing the bell alone (conditioning stimulus). Similarly, Walter regarded the light-seeking capacity of *Machina docilis* as the simple reflex and the whistle as the unrelated stimulus that became coupled to the reflex through training. In further explorations he tested tortoises for which the sound of a whistle initially switched off all motors, rather like the instinctive freezing that many animals show as a startle response. Additional wiring of connections then allowed conditioning of this new response so that the creature could again "learn" that the whistle meant light (and therefore potentially food).

What was this wiring? How did the tortoises work? It was Grey Walter's stated objective to make the electronic circuits controlling the tortoise as simple and as transparent as possible. Built in the pretransistor era, the circuitry was based on vacuum tubes (called valves in the United Kingdom)—glass containers rather like lightbulbs that contain a heated filament. The filament produces a stream of electrons that travels through the evacuated tube to a positively charged electrode at the other end, thereby creating an electric current. Unlike in a lightbulb, however, the stream of electrons in a vacuum tube is made to pass, midway through its journey, through a thin metal screen like a sieve. The screen is connected to other parts of the external circuit and, as it changes voltage, repels or accelerates the electrons passing through the tube. Crucially, a tiny change in voltage on the screen produces large changes in current through the tube. In other words the tube acts as an amplifier (as is true also of present-day transistors).

The small voltages that were amplified by the vacuum tubes came from the two "sense detectors"—the light-sensitive photoelectric cell on the rotating head and the bump detector mounted at the machine's midsection. Thanks to various connections and switches, the voltages from these sensors changed according to the surroundings of the tortoise. If moderate light entered the photocell, the tortoise would move toward it, but it would back off if the intensity rose above a certain value—the dazzle state. Collisions, sensed by the bump detector, changed the circuit from a simple amplifier into an oscillator: the tortoise then moved back and forth in a random fashion until it cleared the obstacle. When an obstacle was encountered in the dark, the avoidance drill was performed in a leisurely fashion, whereas if an attractive light was nearby, the movements became more urgent.

This is in no sense a complicated circuit—two sensors, two transistor equivalents, two relay switches. The series of logical steps it takes, viewed as a flow chart, is about the level of sophistication you might find today in a domestic electric kettle or toothbrush. The design of *Machina docilis,* the robot that was able to learn, was only slightly more elaborate. It had in addition a microphone to detect the training whistle and several more vacuum tubes and switches, allowing signals to be combined with the other sensory and motor activities. But by modern standards this too was trivial. The seemingly goal-directed activity of the tortoise—its ability to learn, its vagaries of motion, its responses to mirrors, its interactions with other tortoises—was generated by short sequences of elementary logical processes. In these terms, Grey Walter's tortoises were significantly less smart than a washing machine!

Might the same be true of an amoeba or a stentor?

The notion that complicated forms and movements can have simple origins is hardly new. A fragment of leaf hanging by a strand of cobweb spins in the darkened corner of a shed; a submerged tree branch ducks and dances in a river flood—objects such as these move rapidly and unpredictably. Their motion obeys the laws of gravity and hydrodynamics, but viewed from a distance or under poor light conditions, they can fool us into thinking that they are alive.

In the late nineteenth century observers drew attention to motile phenomena produced by chemicals. Small fragments of camphor floating on water, for example, skitter here and there like water striders on a pond. Droplets of alcohol on a plate warmed on one side congregate toward that side, droplets of oil move in the opposite direction. Small beads of mercury in dilute nitric acid move in a purposeful manner over long distances toward a crystal of potassium dichromate. Indeed, there was once a theory that the motility of amoebae and other cells is driven by changes in surface tension similar to those displayed in some chemical systems. Sharp-eyed observers such as Herbert Jennings quickly debunked this idea, and nobody today gives it a second thought (although it remains true that interactions between living cells and their substrata are incompletely understood).

I remember a display at the San Francisco Exploratorium consisting of a large rotating three-armed pendulum in the shape of a flattened T mounted on an axle. Each arm had a flexible terminal segment that could swing independently, like a flail. The entire assembly—designed and made by the artist-cum-scientist Ned Kahn—was enclosed in a glass-fronted case. Except that the axle projected through the case to the outside and terminated in a satisfyingly solid metal wheel, like something designed for a submarine or antiaircraft gun. If you gave the wheel a hard twist, the pendulum would spin violently as a unit. But as rotation slowed, one of the three flails would suddenly come out of sync and spin in a different direction. Then the pendulum would hesitate and jerk this way and that—sometimes even reversing completely its direction of rotation. Its movements now seemed completely unpredictable and random, and yet one knew they had to obey the ineluctable laws of physics.

The purpose of the Exploratorium exhibit was to illustrate chaos: intricate, unpredictable outcomes produced by the repeated execution of simple rules. Chaotic systems follow linear sequences of defined steps but are highly sensitive to the starting conditions. Although in principle it is possible to predict their outcome precisely, in practice it is impossible because we cannot define the conditions with sufficient accuracy. Another feature of chaotic systems is that, although unpredictable, they often return to certain patterns of movement. If you play with the three-armed pendulum for long enough, you will come to recognize certain habits, or motifs. There is the energetic carousel, where the entire apparatus spins at high speed; the traffic policeman, with three stationary arms and one solitary flail rotating as if controlling a busy intersection; the conversation piece, with action shifting repeatedly from one side to the other, as though the different parts were engaged in a dialog. The precise beating of the heart, the exact orbit of a planet around the sun are other naturally occurring chaotic systems. They also wander in unpredictable fashion but return ever and again to the same patterns.

So it is easy to understand from a mathematical or computational standpoint how a robot might move in an unpredictable, seemingly motivated fashion. Trace through the wiring of a robot tortoise and ask what happens as this or that voltage rises, and you will uncover a set of rules, such as: PROCEED FORWARD UNLESS OTHERWISE INSTRUCTED; IF YOU ENCOUNTER A LIGHT, STOP ROTATING BUT CONTINUE TO MOVE. The mindless repetition of these rules, like the dynamic interplay between the parts of a three-legged pendulum, is enough to create interesting patterns of movement. This is true even if the environment does not change. If it does change—if any of the external influences such as a source of light or shape of terrain fluctuates in an irregular fashion—the robot's responses will be even harder to predict.

The eminent brain researcher Valentino Braitenberg, in his charming 1987 book *Vehicles,* asked his readers to imagine a series of hypothetical self-operating machines, broadly similar to Grey Walter's tortoises but a notch higher in design complexity. He showed that as sensors, motors, and connections increased in number, so did the range and apparent sophistication of their activities. If you did not already know the principles behind the vehicles' operation, Braitenberg said, you might see in their actions evidence of aggression, love, foresight, and even optimism. It is much more difficult to infer internal structure from the observation of behavior, he concluded, than to create the structure that gives the behavior in the first place.

The set of rules embodied in the wiring of Grey Walter's tortoise are rather like a mathematical formula. You may have encountered puzzles that begin: "Think of a number, multiply it by itself, now subtract 5 from the total" and finish by asking for the final number. This set of instructions can be written down in algebraic form. So if you write x as the number first thought of, then the final answer, y can be represented by a simple equation:

$$y = x^2 - 5$$

The lines of code of a computer program are another example. The progress of a PacMan though its maze will obey encoded instructions of the kind: IF THE RIGHT ARROW KEY IS PRESSED, MOVE TO THE RIGHT . . . IF YOU FIND A BUTTON, REMOVE IT AND ADD 10 TO THE SCORE . . . IF GHOST 2 MOVES ONTO YOUR SPACE, DELETE ONE LIFE . . . IF ONE OR MORE LIVES REMAIN, RETURN TO STARTING POSITION.

My point is that each of these situations represents a kind of computation. In each case, there is a set of instructions, or algorithm, that has to be read in sequence. And there is what you might call a calculator—an electrical circuit, a computer program, or a brain—that acts on those instructions.

The calculator takes one instruction at a time from the algorithm and uses it to change its state. The state of a robot tortoise, for example, is its location in the arena and the direction it is facing at any instant of time. It receives a stream of inputs from the photodetector and bump detector. At regular intervals the robot reads these data and puts them together by means of algorithms in its electronic circuit. After performing what is in effect a calculation, the circuit then activates motors to move forward, to turn, or to back off. So it is when you carry out a piece of mental arithmetic. The algorithm is now the series of instructions to add this, or multiply that. The calculator in this case is your brain, and its "state" is the answer you get after each addition, subtraction or multiplication. Finally, in the game of PacMan the algorithm is the series of movements the player makes with the joystick. The calculator here is the program, and the states are the instantaneous positions of the PacMan, the ghosts, and all the other items within the maze.

In his *Lectures on Computation,* the famous Caltech physicist Richard Feynman talks about the general notion of computation. He speaks of file clerks carrying out simple logical steps by sorting data on cards; arithmetic operations performed with pebbles (think of an abacus); universal computing machines that move along a tape and modify their next action according to what is written on the tape. He even describes a computer based on billiard balls. Balls fired to the center of the table represent the inputs and the distribution of balls coming out of the collision is the output. By adding stationary balls and barriers to act as mirrors to reflect balls, Feynman says that it is possible to perform any conceivable logical operation. In other words, it would be possible to make a computer from billiard balls, although no one would attempt to do such a crazy thing. The physical nature of a computer—the material elements it is made of—is not its most crucial feature. It is the way these elements are connected—the architecture of the machine—that makes all the difference.

The term *computer* also has a more restricted meaning. Modern computers are typically designed around a framework envisaged by the famous Hungarian-born Princeton mathematician John von Neumann, in a paper written in 1945. In this design the instructions, corresponding to the algorithm or program, are kept separate from the part that actually performs the calculations. Instructions from the memory unit are fed together with any necessary data into the processing unit, and part of the resulting output is fed back to the memory unit. From a theoretical standpoint, as von Neumann showed, such a machine is able to perform any conceivable calculation however large or elaborate.

Whether living cells correspond to this closer definition of *computer* is debatable. You might, for example, think of DNA as storing instructions. But with regard to the looser, more colloquial meaning used by Richard Feynman—a computer as a collection of logical elements linked—there is, I think, no question. Living cells qualify. The instructions to a cell are contained in the flood of sensations and stimuli, chemical and physical, it receives from the environment. Its output is its position, orientation, and shape, as well as the position and current activity of its membrane, nucleus, and all of its internal organelles. The calculator is the internal web of biochemical and biophysical processes.

The Grey Walter story has another message, relating to the interaction between humans and robots. His work received wide publicity in the early 1950s and was featured in *Scientific American* and in the tabloid *Daily Express*. Tortoises were exhibited at the Festival of Britain in 1951, and details of their construction, circuitry, and performance eventually appeared in Walter's popular 1953 book *The Living Brain*. Today this work is justifiably regarded as a classic in the development of mobile robots. But if you read Walter's descriptions today you will probably be struck, as I was, by the incomplete and anecdotal nature of his evidence. There are no numbers or statistics to validate his conclusions; no indication how much human intervention was needed to keep the tortoises freely moving within their enclosure.

Walter's interpretations were also highly subjective and ingenuous—even sensationalist. His first article in *Scientific American* begins: "An Imitation of Life: Concerning the author's instructive genus of mechanical tortoises. Although they possess only two sensory organs and two electronic nerve cells, they exhibit 'free will.'" When confronted with a mirror, he writes, the robot begins "flickering, twittering, and jigging like a clumsy Narcissus. The behavior of a creature thus engaged with its own reflection is quite specific, and on a purely empirical basis, if it were observed in an animal, might be accepted as some degree of self-awareness." Elsewhere he makes liberal allusions to "speculation" and "tropisms," "learning" and "hunger," in connection with his tortoises. *Machina speculatrix*, he says, "is never still except when 'feeding'—that is when the batteries are being recharged. Like the restless creatures in a drop of pond water, it bustles around in a series of swooping curves, so that for an hour it will investigate several hundred square feet of ground."

You might criticize Grey Walter for overinterpretation, but he illustrates a common failing. In the novel *The Soul of Anna Klane* by Terrel Miedaner, a woman, Dirksen, is presented with a mobile robot in the form of a small metallic beetle and asked to smash it with a hammer. The robot, like one of Grey Walter's tortoises, is independently mobile and able to locate and feed on a source of electrical energy in the form of a power outlet socket. This robot can also purr like a cat, and it displays a set of small indicator lights that can change color or pulse. Moreover, it is highly sensitive to metal objects, so when Dirksen approaches it with a hammer, it backs away, red lights flashing as though in alarm. It dodges her repeated attempts to smash it until she learns that it has no such reaction to human flesh. Putting the hammer to one side, Dirksen simply picks up the robot (which purrs trustingly) and inverts it onto a bench. As it waggles its wheels helplessly, she brings down the hammer repeatedly onto its body. The creature wails pitifully like a baby, and a pool of red lubricating fluid forms beneath its body.

As this fictional account illustrates, we readily empathize with a suitably designed mechanical toy. With a few more trimmings, a device little more sophisticated than a robot tortoise can evoke feelings of concern and pity that are plainly inappropriate.

Of course, PacMan and mechanical tortoises are ancient history. Our present-day world is full of computer games and animated robots that operate at a far higher level of sophistication. Sometimes the debt to earlier work is obvious. A project by a consortium of European engineers features collections of autonomous mobile robots, Swarm-bots, that act collectively on specific tasks, such as traversing a rugged terrain with hills and chasms or transporting a heavy object. Each robot is a small circular device not unlike one of Grey Walter's tortoises in appearance, equipped with wheels and a flexible arm with a gripper. Swarm-bots carry lights and light sensors; each has a camera and communicates by radio waves. Each robot also carries its own small computer to analyze the movements of its neighbors. It decides on its own actions. Swarm-bots know how to join forces to cross a chasm (using their grippers to self-assemble into a bridge); they know how to pull a large body over a surface (working together like Lilliputians).

The software used to program the Swarm-bots was developed and tested not on real robots but on computer-based simulations, as are used for cars and airplanes. Virtual robots took the place of real, engineered robots in the early stages—a further layer of abstraction. Lines of computer code represented both the electrical circuit of the robot and its movements within a defined arena. This made it far easier and cheaper to try out different designs, make changes to the simulated wiring, and see what effect the changes had on the robot's performance. Only when the design was perfected on the computer did the designers move to the far more lengthy and expensive process of actually making the robot.

Our readiness to interact with robots has been exploited by the toy industry, and a range of electronic pets can be purchased for the price of a small computer. For example, you can buy a sleek dog, AIBO (for Artificial Intelligence robot), in black or white pearlescent plastic, offering a simulated pet experience to a caring owner. AIBO can walk, play soccer with a ball, chew a bone, sit, lie down, roll over, and feed itself at a power outlet. It can emit doglike sounds and express a variety of emotions and instincts using an illuminated face and mobile tail and ears. The animal can recognize up to three owners by their voices and will learn to identify favorite locations through built-in image-analysis software. An additional feature (eerily appropriate to the theme of this book) allows the pet owner to, in a sense, enter the dog's head. Through a video camera mounted in the front of the head, the user can view the world as seen from the dog's eyes.

Perhaps the most appealing (or insidious, depending on your view) feature of AIBO is its capacity to develop a distinct character. It begins like a puppy, learning how to move around and to recognize and pay attention to its owner. Tactile sensors on its back, head, and chin trigger realistic responses to petting and grooming and discipline. The animal simulates such emotional states as pleasure, surprise, and anger through its external lights, moving parts, and sound production. No wonder a child playing with such a creature on a daily basis soon comes to think of it as a real pet.

Software advances since the early PacMan days have been astounding. Virtual reality displays, computer games, and immersive synthetic worlds have attained an undreamt of sophistication and penetration. Perhaps the single most significant advance from the perspective of this book has been the introduction of avatars. This Hindu term refers to the manifestation of a deity, especially Vishnu, in human, superhuman, or animal form. In today's parlance, it refers to a movable image on a computer screen representing a person in a virtual reality environment. Avatars represent a major step along the road of human identification with virtual images that started, in this chapter at least, with PacMan. Instead of being cryptic and implied, the assignments here are deliberate and declared. The virtual reality avatar takes your part in a variety of situations; it often carries your personal characteristics. It is your alter ego . . . it is *you*.

Adventure and war games, sometimes referred to as FPS, for first-person shooting games, feature avatars in hazardous environments battling to achieve target aims. The figure of a marine in camouflage gear leads the way through a war zone, aiming to capture an enemy position in an abandoned factory. Crouching low, she dodges bullets, seeking protection where she can. A hero without question, she leads the way through the chaotic war zone of Baghdad or Beirut, signaling back to her buddies when she finds safety.

The avatar in this case is an off-the-shelf, generic soldier. Her viewpoint may be either first-person—what the warrior has in her field of vision as she turns her head, the player sees on the screen—or bird's-eye, with the camera floating above the avatar. Either way, the avatar moves in response to the keyboard or joystick instructions of the user, who easily becomes immersed in the synthetic environment. Online versions of FPS have different avatars battling in real time—twenty-first-century connection speeds allow dozens of players to participate with split-second timing. The player's initial feeling that "my character is performing on my behalf" quickly evolves to a sense of "I am there." The games are also populated by other figures that are not avatars (no real person pulls their strings) but animated cartoons. Some are well-known personalities in their own right, such as the energetic, voluptuous Lara

Croft, star of the Tomb Raider adventure games and sex symbol for a generation of adolescent boys.

The fastest-growing sector of the computer-generated-games industry is that of massively multiplayer online role-playing games, or MMORPGs ("morpegs"). *Massively,* for purposes of this acronym, refers to a game with at least one million subscribers. At the time of writing it is estimated that an astonishing one hundred million people worldwide spend significant fractions of their day immersed in these virtual environments. Some MMORPGs offer Tolkienesque fantasy worlds where players do battle with dragons, quest for fairy gold or magic potions, and build up their virtual wealth and power. Others have a more amorphous social context, allowing avatars to work, play, and live with one another in cyberspace, a sphere of socialization distinct from everyday life. Surveys indicate that a significant proportion of players regard MMORPG worlds preferable to their own—and even more real.

Perhaps the most curious feature of MMORPGs, in view of their evident ability to engage the unwavering attention of millions, is that they are not particularly convincing from a visual or sensory standpoint. Yes, they have land and oceans, cities and forests, beautiful people. You can find games in which sunlight glints on waves, the moon peeks through branches, enormous skyscrapers reach up forever into the sky, tier after dazzling tier. But a casual passerby would never for an instant be deceived into thinking that the screen was a window onto an actual new world. Buildings, rock formations, grass, and vegetation are all rendered with a broad brush and lack the proliferation of details and imperfections that characterize our daily experience. A typical avatar in a multiuser game is a coarse-grained cartoonlike image with a restricted set of sticklike actions and expressions. Your communication with your avatar is limited to typing messages on a keyboard such as PICK UP LOGS, BUILD FIRE, or ENTER ROOM.

That such crude and clunky interfaces are sufficient to engross players testifies to the power of role playing. We willingly suspend disbelief in order to become immersed in the new world. Greater realism is evidently not necessary in these games and might indeed be difficult to achieve for a game with many millions of participants. But for individuals the story is quite different. The technology already exists, and has for several decades, that can delude us into thinking we are somewhere else.

Science fiction writers and other visionaries have long anticipated the consequences of sensory inundation. Aldous Huxley's *Brave New World,* written in 1932, just a few years after talking motion pictures became widely accessible, envisaged a form of cinema in which the audience received odorous and tactile sensations in addition to visual and auditory. By grasping metal knobs on the handles of their seats they receive a stream of sensory input, experiencing a love scene on a bear rug between a black man and a pneumatic blonde in vivid detail, each hair felt with prickling reality. Actual virtual reality devices followed slowly and at some distance: for example, a device known as a Sensorama, constructed in the 1950s, simulated the experience of riding a motorcycle. The participant sat on a pillion seat holding mockup handles and faced a display of an oncoming road on a rather small cathode-ray tube. By the early 1960s there were head-mounted virtual reality displays that the user wore like a mask. Wands and gloves relayed data about the individual's arm and hand movements to a computer.

Military applications of these devices were quickly realized. The radar developed in the Second World War with its cathode-ray tube became the outside world for a generation of pilots and navigators. Flight simulators became increasingly realistic, incorporating not only visual displays but also physical rocking and rotation that conveyed to the operator the sense of movement. The human brain, it was quickly learned, is easily deceived about the degree and magnitude of movements. Even small accelerations and changes in tilt and altitude, when coupled with appropriate visual stimuli, give a dramatic sense of rapid movements. In a 1992 incarnation of these ideas, developed at the University of Chicago, the subject enters a small cubicle and is surrounded by visual and auditory stimuli. Similar CAVE installations—the name inspired by Plato's famous analogy of physical reality with shadows cast by firelight on the wall of a cave—now find numerous commercial applications as in the design of naval ships and automobiles.

To a degree, virtual reality enters into our everyday experience. Everyone has used tools and devices that extend our bodily capabilities. You have no doubt experienced, at least in a small way, the sense that instruments are part of your own body. Users of microscopes and telescopes, operators of cranes and mechanical arms, even automobile drivers all have, from time to time, felt a sense of identity with their machines. Workers in facilities who handle radioactive materials through feedback-controlled mechanical arms and hands routinely report a shift in point of view. They recount that they sense through their metal arms and fingers the weight and slipperiness of the metal cylinders. They know, of course, that the sensation is an illusion. But that does not stop them from feeling as if they are inside the isolation chamber, among the items of equipment they are manipulating.

It has been known since antiquity that arms or legs torn off in an accident or removed surgically often continue to signal their presence. These phantoms were named and studied in great detail by the American neurologist Weir Mitchell during and after the American Civil War. Phantom limbs seem to remain attached to the body even though no neural connections remain; some are ephemeral and ghostlike, whereas others are astonishingly real. Amputees can feel the entire arm or leg and can "move" it, sometimes reporting that they can wriggle their nonexistent fingers and grasp objects. Perhaps the most famous phantom limb belonged to Admiral Lord Nelson, whose right elbow was shattered by a musket ball at the battle of Santa Cruz in 1797 (part of the French Revolutionary Wars) and who subsequently lost his lower arm. Nelson continued to feel the presence of his right arm long after, with such semblance of reality that he took it as evidence of an immortal soul. After all, he argued, if a single arm can persist after being lost, then why not the entire body?

The neurologist and brain scientist V. S. Ramachandran made a study of phantom-limb syndrome and uncovered a remarkable series of phenomena. In at least some of his patients the sensation of an arm or a leg being present after amputation seems to have been due to rewiring in the brain. That is, following the loss of a limb, areas of sensory cortex that normally receive input from muscle spindles and skin receptors in the limb receive sensory input from other regions of the body. The areas most likely to substitute are those whose sensory projections are anatomically closest to the limb in the normal brain mapping—the human sensory cortex has a distorted body plan, or homunculus, that maps inputs from the body. Ramachandran and his colleagues found that patients who had lost their right arms, for example, often reported sensations in that phantom limb when their right cheeks were stroked, the mapping of arms and face being closely located in the sensory cortex. In some there was a distinct representation of individual fingers on the cheek and jaw. The modality of the stimulus—whether vibration or stroke, wet or dry, hot or cold—was also preserved in the displaced sensation.

The experience of a phantom limb can take a fascinating variety of forms. In some patients the illusion is long lasting and colored by past memories—retaining the impression of a wedding ring, for example, or a wristwatch. In others the illusion changes over time: the arm might shrink in length until the hand and fingers seem to be attached at the shoulder. Ramachandran and his colleagues also found that visual stimuli could modify the sensations. One of his patients experienced persistent pain owing to what he felt as an uncontrollable clenching of his fist (with nails digging into his palm). When this patient was presented with a mirror reflection of his good arm, appearing in the location of the nonexistent phantom, he found it at last possible to unclench his fist. For the first time in months he experienced relief from his pain.

What do these studies tell us? Apparently, our sense of the shape and extent of our body—what the English neurologist Sir Henry Head termed our *body image*—is malleable. It can change, even in adulthood. If you have the misfortune to lose an arm or a leg then, chances are, you will continue to feel the missing limb, at least for some time. You might experience pain from the limb or have the sense that it moves around (perhaps gesturing as you speak). Illusions of an opposite nature can arise when the damage occurs more centrally—not to the limb but to the brain. A victim of stroke whose right temporal lobe is affected frequently suffers paralysis of the left side and sometimes has radically altered images of the outside world. Otherwise rational patients might deny that they are in fact paralyzed on one side, and go to great lengths to explain away the obvious. They may ignore or even deny the existence of their left side. In his essay "A Leg to Stand On," the neurologist Oliver Sacks describes the case of a young man admitted to hospital with a "lazy" left leg. This patient fell asleep in the hospital bed and then awoke with the discovery, as he perceived it, that someone had put a severed leg in the bed. Trying to throw out the disgusting object, he found that it was somehow attached to his body. Nurses found him on the floor beside the bed, still furiously trying to remove the offending limb. Despite the clear evidence of his own eyes, he believed that the leg did not belong to him. His own left leg, which he knew intellectually must exist, was in some fashion lost.

The message is clear: there is no rigidly defined, hard-wired set of connections between sensory inputs coming from the arms and legs and their representation elsewhere in the brain. Although innate specifications do exist (individuals born without arms sometimes "feel" these limbs to be present) this representation is malleable and readily altered. It can be changed not only by traumatic events such as injury or disease but also by exposure to hallucinatory drugs or even in relatively mild psychophysical experiments. You or I can be tricked into feeling that our nose has become stretched to inordinate lengths, or that a dummy hand has taken over from our real hand. In a well-known experiment, each subject was positioned in a hospital magnetic resonance imaging (MRI) machine with her right hand hidden beneath a table and a realistic rubber hand visible on the tabletop. A researcher used a small brush to stroke the finger of the hidden real hand while simultaneously stroking the corresponding part of the rubber hand that the subject could see. Within fifteen seconds the subject typically developed a profound sense that the rubber hand was part of her . . . her own hand. She flinched when the dummy hand was threatened with a hammer and expressed frustration when she discovered that she was unable to move a rubber finger. Although she *knew* intellectually what was going on, she was unable to dispel the sensory illusion. Brain scans from the MRI suggested that the portion of the brain involved in the illusion was one known to integrate vision and body movements.

Once again, these experiments tell us that our body image is a construct, continually revised and updated. It is rather like a hypothesis, built from the available sensory data and installed into its correct place in our view of the world. Sensations such as pain are an opinion about the state of the body. We are, as Prospero told us, such stuff as dreams are made on.

But I've drifted a long way from the central theme of this book. What possible connection can there be among phantom limbs, mechanical arms, computer games, and crawling cells? In short, we humans perceive these disparate phenomena by the same biological apparatus—the brain. Each image is exposed to the same subjective filtering. Our experience of our immediate surroundings is, like that of our body image, a hypothesis. Inundated by sensory inputs of different modalities every second of my waking day, my brain has to stay afloat. It has to make sense of this flood of inputs in order to take decisions regarding future actions. How it does so is invisible to me, the mechanism operating at an unconscious level. But its conclusions seem utterly convincing. Indeed, the very fact that I formulate these ideas by an invisible path no doubt explains their innate quality. In a similar way, I suggest, my interpretation of objects that move in a certain way, such as pixelated forms on a screen or cells viewed in a microscope, is also colored by inner prejudice. In this case the input is solely visual (or visual and auditory) and the interpretative role of the brain relatively innocuous. But my tendency to attribute sentience to crawling cells or even computer animations flows from the same source as my sense of our limbs and our body image.

I assign purpose and sentience to moving objects instinctively because of the way my brain is built. In the environment of a jungle or savanna this assumption is the right one to make. A form that moves with stealth toward me or that runs away as soon as I am within sight is certain to be living. Its motion flags a creature with immediate significance for my survival. So when I observe an amoeba crawl in pursuit of a microbial prey, I unconsciously apply an association embedded in my brain by evolution. I assign to that image the attributes of a sentient organism. When I watch a paramecium dodging here and there in search of food, I am congenitally unable to assess its inner state, even though I know intellectually that on the basis of behavior alone it might be no more complicated than a PacMan or a robot tortoise.

Then again, it could be much more. Recall that the microorganism is indubitably alive. It was not designed by humans to move like a toy but arose spontaneously (albeit over a long period of time) out of nonliving materials. Visible movement is just one of its attributes, and not the most vital. You may laugh at the idea that an amoeba or a paramecium has feelings or a humanlike consciousness. But it seems equally improbable that there is nothing more to a migrating cell than a few hundred lines of code in a PacMan program. Even if decisions to turn this way or that are made by a relatively small set of molecules, this decision apparatus cannot exist independently of the cell. There has to be an underlying infrastructure of other molecules that has evolved to support this motion. The very existence of a cell, its ability to feed and reproduce, to regulate its uptake of nutrients, is a massive achievement. These other aspects of life do not create immediately observable movements that can be seen under a microscope. But they contain a world of sophistication and intelligence of a kind that we find much harder to recognize.

Just as we overestimate the sophistication of things that move, we *under*estimate those that do not. If it doesn't move, we unconsciously assume that it must be stupid. Of course, this must be a fallacy. The laptop computer on my desk performs sophisticated calculations and logical operations at a speed I could not possibly match. It easily beats me at chess. But it also sits there, a cold slab of metal, like a dumb piece of furniture. Similarly, we forget, as we play a game such as PacMan, the highly structured algorithmic framework for the graphical movements. Someone had to make the power pack and supply the electricity; someone else had to design the case and the computer screen. Teams of anonymous technicians spent years putting together the underlying circuitry of the operating system and writing programs that place pixels at specified positions.

You can say the same about Grey Walter's tortoise. Imagine taking one back in time and placing it on the floor of a Stone Age cave. What would the inhabitants of the cave make of such a creature? My guess is that they would find it frightening and magical, not just because of its movements but because of its very existence. That assemblage of metal and glass parts with its alien anatomy and independent motility would have seemed incomprehensible—a creature of the gods. And in a sense those cave men would have been correct. A robotic tortoise embodies thousands of years of culture and acquisition of knowledge. We take it for granted because we live in a world full of such objects. Only by seeing it through the eyes of an outsider can we appreciate how wonderful it is.

The same must be true of single-celled organisms, only more so. If you regard their movements as no more than a simple series of logical decisions, then they are indeed trivial. Computer programs that reproduce essential features of their movements and responses are easily written. But these movements are just ripples on a very deep pond; diving below the surface, we will find a world of molecules, working ceaselessly to provide a suitable framework. Because this infrastructure is invisible to a casual observer, its sophistication is consequently discounted.

Like a robot tortoise in a Stone Age cave, the very existence of a motile free-living cell implies a long history of development and a universe of hidden machinery. Most remarkable of all is that a bacterium or a stentor is capable of self-assembly: it actually makes all of the equipment needed for movement as it grows and divides. It is as though a robot tortoise could gobble up particles of glass, metal, and silicon and, after a period of gestation, give birth to a new, functioning robot. There is evidently far more going on inside a living cell than you can see just by looking down a microscope. Its external movements tell you next to nothing about its internal states and logical intricacy . . . how much or little it knows of itself.

THREE

Protein Switches

A good place to start our descent to the realm of the very small is with thermal diffusion. This fundamental physical process dominates atoms and molecules and is crucial for all aspects of living cells. It is also an essential ingredient of wetware, providing the equivalent of connections, or wires, for the cell computer. Diffusion can be seen directly under a microscope and is something every biology student discovers afresh. You are examining tissue cells flattened on the surface of a plastic culture dish, or a drop of saliva on a slide. As you move the focus control up a little into the fluid you suddenly see it: a minute speck of matter jiggling endlessly in place. How fast it moves depends on size: large particles barely bestir themselves, whereas the smallest particles skitter about like mayflies. The best way to see the movements is to illuminate the specimen from the side. Particles now appear as spots of light, darting here and there against an inky black background.

The botanist Robert Brown first described these movements in 1827 when he observed minute particles dancing within fluid spaces in pollen grains. He later confirmed the presence of similar "active molecules" in a variety of samples prepared from sulfur, resin, and wax, thereby demonstrating that the particles did not have to be alive. The origins of their perpetual motion, however, remained a matter of debate until Albert Einstein provided the correct physical explanation. In a paper published in 1905, based on his doctoral dissertation to the University of Zurich, Einstein argued that Brownian motion was due to the continual agitation of the molecules of water. Shortly afterward, the French physicist Jean Perrin validated Einstein's theory experimentally. His measurements provided one of the first estimates of the size and speed of motion of molecules.

The molecules of a fluid move continually at high speeds, colliding with each other frequently so that they are continually knocked off course. Their movements are due to heat energy, and their average speed is directly related to the temperature. At very low temperatures the molecules of water hardly move, being locked into the rigid structure of ice. But as the temperature rises, their agitation increases and they break free to form first a liquid and then, if the heating continues, a gas. A large particle suspended in water at room temperature experiences many tiny impacts at every instant, but their effects are relatively small and mostly cancel out: the particle barely stirs. But for a particle the size of a bacterium or smaller, the effects are far greater, and the particle is displaced this way and that. For an even smaller particle, say the size of a protein molecule, the motion becomes insanely vigorous. Dancing violently, the molecule skips along a meandering path from its starting point. The result is diffusion.

The interior of living cells is a strange environment—rather like a chunky soup or semisolid slurry. There are membranes and large particles of all sizes, including the nucleus, and the bacterium-sized mitochondria that generate energy. If these large structures move at all, it is with a relatively slow and orderly motion driven by molecular reactions. But the watery contents of the cell, the dissolved salts and unattached molecules, are in continual motion. They are constantly stirred by thermal diffusion. If you introduce a fluorescent marker in one region of a cell and then watch it under a special microscope, it will instantly bleed into the rest. A small molecule such as a sugar can cross from one side of a human cell to the other in a tenth of a second; a protein might make the same journey in a second. Diffusion is a very effective way to move over small distances.

Because of thermal diffusion, any two molecules in a cell have a good chance of meeting each other within a short space of time. Although their encounter is usually fleeting and inconsequential, it can lead to a meaningful change. For example, molecules might agglomerate into a larger body. Or if one molecule is an enzyme (a protein that performs a specific reaction), it might provoke a chemical change in the other molecule. From the standpoint of cellular computations, diffusion provides the connections. It is a miniature World Wide Web that allows every molecule to address every other molecule in a short space of time.

But if atoms are continually buffeted by thermal energy, why doesn't a cell burst apart or explode into a puff of gas? The answer is that certain kinds of atoms like to stick to each other. Chemical bonds resist the dissipative forces of thermal energy and hold atoms close together for long periods of time. The agents of chemical bonding are the outer electrons of each atom. Why these bonds form, when, and how strong they are, are questions for chemists. But for my present purposes I need only distinguish two kinds: strong and weak.

Strong bonds are those that hold fast for long periods at the ambient temperature of a living cell—the cement that holds it together. Sugars, amino acids, lipids, nucleotides, RNA, DNA, and proteins are all built from chains and rings of carbon atoms linked to each other by strong bonds. Once made, strong bonds are usually stable. But they can be made and broken down by biochemical processes, guided along appropriate pathways by enzymes. Their formation is driven by what is called chemical energy: the irresistible combining power shown when certain molecules come into contact. Chemical energy drives many kinds of processes: the explosion of gunpowder, the burning of coal, the movement of pistons in an automobile engine, and the contraction of our muscles.

People in the affluent West eat so well and so often that the connection between food and energy is often lost. But under conditions of starvation, the relation is obvious. For the inhabitants of the Soviet labor camps, described by Aleksandr Solzhenitsyn in his harrowing *One Day in the Life of Ivan Denisovich,* two hundred grams of bread was the equivalent of a day's work, twenty-four hours of survival in the deadly cold. The basis of this stark equation is that molecules of bread contain chemical energy. This energy, derived ultimately from sunlight and captured by plants, is released in our body as we eat. Precisely how that happens is complicated and entails many different biochemical reactions, but in broad terms it is a form of combustion. When molecules of bread combine with oxygen brought into the body by breathing, the energy is released. But instead of being lost entirely in the form of heat, as it would be if the bread were put into a fire, the chemical energy of bread is released piecemeal. Each parcel of energy drives a particular chemical reaction catalyzed by an enzyme. These reactions include all the processes that keep us alive: movements, mental activity, thermal regulation, repair, and maintenance.

The energy parcels to which I just referred are small molecules specialized to carry chemical energy. They scurry from reaction to reaction inside the cell, picking up energy from one and using it to drive another. One energy carrier stands out from all the rest by virtue of its abundance and widespread use—adenosine triphosphate (ATP). This small molecule has a tail made of three phosphate groups (a phosphate being simply a small group of atoms containing one phosphorous and three oxygen atoms). The terminal phosphate in ATP requires an input of energy to be formed but can release energy when it is lost again. It is sometimes referred to as being in a "high-energy state," as though spring-loaded. Energy available in food molecules is used to make ATP—winding up the spring, so to speak. Transfer of the phosphate to another molecule releases the spring of ATP but winds up the second molecule. The receiving molecule is then able to react with yet another molecule.

Let's assume that an amoeba has just engulfed some food. Digestive enzymes degrade the large molecules of the prey into easily assimilated sugars, amino acids, and lipids. The amoeba uses a portion of this harvest as metabolic fuel to drive the formation of more ATP. The rest provides building blocks for RNA, DNA, and proteins. Manufacturing these amoeba-specific substances means that strong bonds have to be made; ATP provides the necessary driving force.

Weak bonds are those that are easily broken by thermal energy at the temperature of the cell. They are created when two diffusing molecules, meeting by chance, find that their molecular shapes mesh. Small interactions between the two faces then hold the two molecules together, their number and strength depending on the extent of overlap of the two molecules and how well their surfaces match. If the surfaces dovetail perfectly, many weak bonds are formed and the two molecules might stay together for the life of the cell. But if there is only a partial matching, only a few bonds can be formed. Those two molecules will not stay together for very long before being torn apart by thermal energy.

Weak bonds enable one molecule to recognize another and will be especially relevant when I talk about proteins. Proteins are made as very long chains of simpler molecules, amino acids. Folding in a precise manner, the chains create convoluted surfaces: patterns of atoms that they present to the outside world and that govern their interaction with other molecules. The powerful lock-and-key metaphor proposed in 1895 by the German chemist Emil Fischer is useful here. This depicts the interaction between a large molecule such as a protein and a smaller molecule such as a sugar as a fit between two complementary surfaces. If the larger molecule has a cavity on its surface that exactly matches the smaller one, then the two should come together and stick. Other molecules (which have different shapes and consequently fit less well if at all into the binding pocket) will be excluded from the marriage. Even today, molecular biologists use the lock-and-key metaphor to explain how enzymes interact specifically with their target molecules. It is how antibodies—the proteins responsible for our immunity to foreign organisms—can pick out just one type of molecule out of billions of different forms. Biologists use the term *recognition* for this process.

And, to muddle the metaphor, weak bonds are the glue that holds lock and key together. Sensitive to the smallest changes in structure, the subtlest nuances of chemistry, weak bonds measure the compatibility of two molecules, test the goodness-of-fit. Molecules are brought together in a myriad fleeting encounters every second by the unending riot of thermal motion. Weak bonds between their surfaces determine whether they stay or go: they arbitrate whether the encounter will become lasting.

Complicated and yet precisely defined, proteins are amazing molecules. *Hemoglobin,* the oxygen-carrying molecule from blood, was the first protein to be crystallized. It was the first (together with its close relative myoglobin) to have its atomic structure determined. Human hemoglobin is built from 574 amino acids arranged in four chains that loop and fold back on themselves like spaghetti on a plate. But although they appear unruly and disordered the chains are in fact arranged with amazing precision. They have the same folds and their amino acids are in exactly the same positions, oriented in exactly the same direction, in molecule after molecule after molecule. That this is so is one of the marvels of evolution, for if you made a protein from an arbitrary sequence of 574 amino acids, it would probably produce glop—an amorphous and undefined coagulate. But over billions of years, natural selection discovered rare sequences of amino acids that fold stably into a defined shape—a different one for each protein.

Proteins such as hemoglobin are large enough to have an inside and an outside. Regions of the folded chains buried near the interior see only other amino acids. They dovetail with their neighbors, or stick to them like flies to flypaper. Their role in life is chiefly to hold the large molecule together, stably and in place. But amino acids on the outside are exposed to the big, wide world. Their precisely arranged atoms create convexities and concavities, patterns of atoms and electrical charges on the protein's surface. They create an intricate chemical landscape: one that other molecules encounter and, in some cases, recognize through lock-and-key matching.

The primary function of hemoglobin is to bind oxygen, and the amino acids on its surface achieve this purpose. Each of its four chains has a special nook where a single molecule of gaseous oxygen fits; this is how hemoglobin carries oxygen throughout our body. But for enzymes, binding is only half of the story. Their evolution has carried them to a

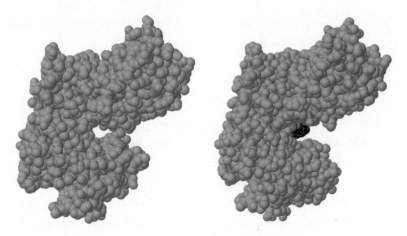

FIGURE 3.1. A protein molecule. The enzyme hexokinase is shown with individual atoms on its surface represented as small spheres. The enzyme has a deep cavity into which the small molecule glucose fits—as shown in the right-hand image.

seemingly higher level of sophistication, where they produce a chemical change in their partner. An enzyme leads its bound molecule (its substrate) through a sequence of chemical steps with the single-minded intensity of a tango instructor. Each sequence is different, each enzyme reaction chemically unique. Some enzymes split their substrate into two pieces, usually with the addition of water molecules; other enzymes join two substrates to make a large molecule (removing water in the process). There are enzymes that add atoms to other proteins; enzymes that cut DNA, fold it, and tie it into knots.

So you have to imagine cells full of billions of protein molecules, of diverse shapes and sizes, continually twisting, turning, vibrating, and moving past each other. Each kind of protein has a unique chemical surface that governs its social interactions with other molecules. It can recognize other specific molecules through diffusion and binding. But there is one other property that is crucial for my developing argument: proteins can exist in more than one state, a discovery made more than forty years ago.

In the summer of 1961 two hundred or so biologists assembled at the Cold Spring Harbor Laboratories on Long Island, New York. Their meeting was entitled "Cellular Regulation Mechanisms," which even to the participants must have seemed dry. After all, *regulation* usually means uniformity and lack of change. But from a historical perspective the outcome of the meeting was a clap of intellectual thunder. At the end of the eight days, the French biologists François Jacob and Jacques Monod presented a summary that would eventually illuminate molecular events in every interstice of a living cell. Most important, in the context of this book, they showed how in principle living cells perform computations.

The novel idea that had bubbled up independently in several laboratories and now found a unified voice was that proteins are able to recognize more than one molecular partner—that is, that they have *two* binding sites. These two sites, moreover, seemed able to interact with each other, talk to each other from one side of the molecule to the other. The simple step from one recognition site to two seems innocuous at first, but in reality it was a revolution. Why? Because it meant that a protein involved in one cellular process—for example, an enzyme making glucose—could be functionally linked to another process through its second site. And this second process could be entirely unrelated. Yes, it might be a reaction involving another sugar. But it could equally well be the copying of DNA, the synthesis of testosterone, or the manufacture of an eye pigment.

The deep message was that protein molecules allow evolution to "wire" the reactions of a cell in whatever way it pleases.

But how can different parts of a protein molecule talk to one another? Once again, hemoglobin provides the answer. Red blood cells crammed full of hemoglobin molecules circulate within the blood. They pick up oxygen in the lungs and release it in the fine capillaries elsewhere in the body. If you were to ride with an individual hemoglobin molecule on this route, you would see it functioning like a lung in miniature. As it binds oxygen, the protein expands slightly, opening out to a more relaxed and flexible form. This is the bright crimson oxyhemoglobin of arterial blood. Later, after hemoglobin has delivered oxygen to the brain, muscles, or kidney, its shape changes in the opposite direction, becoming the tightly knotted, dark red deoxyhemoglobin of the veins. Elucidation of the atomic movements underlying these changes in the 1960s showed for the first time subtle changes in different parts of a protein molecule. Adding an oxygen molecule to the hemoglobin molecule causes some of the amino acids to shift in position by a minute amount. One amino acid influences another through weak bonds leading to a cascade of changes through the molecule.

In simple terms, the hemoglobin molecule is a molecular switch: it flips from one shape to another when it binds oxygen. It does so in order to transport oxygen. But that hardly explains why other kinds of protein should also have a Jekyll and Hyde character. Back in the 1950s biochemists were like children exploring a huge box of toys. As they dug deeper into the wonderful world of enzymes, the reactions they discovered were amazingly versatile and surprisingly subtle. Individual reactions in a cell rarely proceed in an unrestrained fashion. Hundreds of enzyme reactions that make and degrade small molecules spend as much time standing still as they do in motion. The enzyme that catalyzes the first step in the sequence of reactions that synthesize an amino acid, for example, works at top speed only if that small molecule is in short supply. When the amino acid is plentiful and the cell does not need to make more, the enzyme simply shuts down.

Control also exists at the genetic level, in the manufacture of enzymes. The hereditary information of cells that specifies the structure of a protein is contained in DNA. This molecule consists of two very long chains of four kinds of nucleotide bases (small, nitrogen-rich molecules) linked in a precise sequence. The instructions for each protein are contained in a discrete region of this sequence called a gene. Other proteins bind to DNA and control whether a particular protein is actually made (expressed). For example, a protein called a repressor binds to the genes that allow bacteria to use the sugar lactose. If no lactose is available, then the repressor simply clamps onto DNA and prevents the genes from being made. However, if lactose enters the cell, then it binds to the repressor and switches it to an inactive state. The clamp is lifted from the DNA, and the cell manufactures the enzymes necessary to use the lactose as food.

These two types of regulation, one enzymatic and the other genetic, both serve the economy of the cell. But until the famous meeting at Cold Spring Harbor, it had not been appreciated that they also share a common mechanism. What Jacob and Monod demonstrated was that both phenomena depend on proteins that recognize two distinct molecules. An enzyme catalyzes a chemical conversion of one amino acid unless it sees another amino acid first; a repressor protein sticks tightly to a specific region of DNA unless it also happens to encounter a sugar. In both cases there is a primary target and a secondary regulator, and these two molecules have no necessary chemical relationship to each other. To emphasize this feature, Monod and Jacob coined the word *allostery,* from Greek words meaning *other* and *shape.*

Since the primary target and the regulator recognize different locations on the protein surface, their binding will have different effects. A regulator, in particular, will always bind in such a way as to promote a change in the shape of the protein molecule. Small shifts in atomic positions propagate through the structure and modulate the shape of the primary site, making it more or less open for business. Often a small molecule will bind to and switch off an enzyme. This is found, for example, when there is a chain of enzymes that convert some starting material into a product, working like the different stages in a factory line. It is important in these circumstances that the cell be able to control the amount of material that enters the production line; otherwise, it might make too little or too much product. An almost universal strategy found in cells, known as feedback inhibition, is that the end product acts as a regulator of the first enzyme. If the final product is in short supply, the enzymes work full out. But if the product builds up, so that the cell has a glut, then it inhibits the first enzyme, and the factory line slows down accordingly. The same logic, incidentally, operates in a thermostat when a rise in temperature switches off a room heater.

Two-state, allosteric proteins are everywhere in a living cell, in every nook and cranny. They are the building blocks of flagella and cilia on the surface of a cell and of the numerous filaments that crisscross its interior. They are the molecular motors that harness chemical energy to make a cell divide, a muscle contract, and, yes, an amoeba crawl. Two-state proteins embedded in cell membranes open and shut to allow ions and small molecules to cross into and out of a cell. They are at the heart of the electrical signals produced by nerve cells in our brain and hence underpin all our mental activities. Two-state proteins control the copying of the genetic material DNA and the manufacture of proteins. Acting as switches, they respond to signals from elsewhere in the cell and hence determine whether particular genes are on or off.

As Jacob and Monod realized fifty years ago, proteins are conditional connections between different processes of a cell. They are molecular switches. One connection reads cell energy levels by binding to a specific small molecule and, as required, triggers the release of stored energy. Another protein links the synthesis of an enzyme to the presence of a key hormone. Yet another opens a water channel in the membrane to allow the concentration of salts in the cell to be adjusted. At the heart of every one of these causal links is a protein that recognizes two cellular processes. A change in one process causes the protein to switch its shape; in this new state, the protein modifies a second process.

My argument goes as follows. Living cells must be capable of some sort of logical analysis or they would never survive. How else could

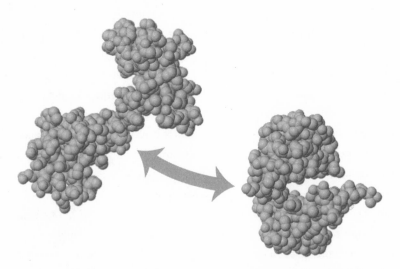

FIGURE 3.2. A protein switch. The calcium-binding protein calmodulin undergoes an especially substantial change in shape.

free-living microorganisms avoid obstacles, hunt for food, respond to stimuli, store memories? If you saw a mouse or a monkey doing these things, you might say "Well, the animal has a nervous system," meaning that it contains a network of interconnected nerve cells and muscles, linked in some complicated fashion in the brain. But a single cell has no nerve cells, so something else must be doing their job. And the obvious candidates are protein molecules. They are, to begin with, the most abundant material of a cell: cells are 70 percent water, but more than half the remainder is protein. Protein molecules have amazingly diverse, intricate structures and perform incredibly sophisticated chemical tasks. What is more, as just mentioned, they have the ability to exist in two forms: two shapes that can have totally different properties. Flipping between these two states, in a manner sensitive to their surroundings, they perform like molecular switches.

Imagine yourself as an enzyme molecule in a cell. Because of thermal agitation you are bouncing around with insane agitation, jiggling so violently that only your strong bonds hold you together. This same thermal motion causes other molecules to bombard you from all sides, hitting you a trillion times every second.

Now the vast majority of collisions are fleeting and inconsequential, but not all. You are programmed to recognize one particular, rare molecular species in the crowd milling around you. Let's call this special molecule substance A. Your task as an enzyme is to fish out molecules of A from the weltering crowd. So when, at last, one does happen to blunder into you, events take a different course. Instead of the usual casual bump-and-bounce, this encounter becomes slower and more lingering. Convexities match concavities, positive and negative charges match, electrical changes neutralize, oily patches stick. Molecule A buries itself into a precise spot on your flank.

Now you respond. Your chains of amino acids shift slightly; a water molecule you have been nursing for just this eventuality pushes up against the visitor. The A molecule in your charge becomes deformed by these motions. It bends, some of its electrons become displaced, and . . . snap! The act is consummated. The A molecule in your grasp is transformed into a chemically different molecule, call it A' ("A prime"). Perhaps it lost some atoms, or acquired some new ones, or it may have become changed in some other way. But whatever the case, A' has a different structure from A. No longer fitting snugly into your active site, it immediately diffuses away and becomes lost in the crowd. You promptly set off fishing once more.

In the world of atoms and molecules, such events are extremely fast. A typical enzyme will convert substrate to product (A to A') in a millisecond or less. Some superenzymes do it even faster, converting millions of molecules every second.

From the standpoint of cell chemistry you are a conduit, a one-way pipe, taking up molecules of A and spitting out molecules of A'. How quickly you do this obviously depends on supplies. If A molecules are scarce and hard to find, A' molecules will dribble out of the pipe one at a time. If A is abundant, the flow rate will become faster and faster—up to some limit.

You have still not performed any calculations, but this is where your Jekyll and Hyde character comes in. You, the enzyme, can adopt two different shapes. In one of these (call it your "on" state) you convert A to A' as just described. In the other, "off," state you are inert—unable to function as an enzyme. And how are these two states controlled? Why, by the binding of a second molecule, call it B. If B is absent, you are off, inactive: the pipeline conversion to A' cannot proceed. But if you meet and bind a B molecule, you flip to the on state and the reaction can occur. In terms of logical operations you are performing the following:

$$\text{IF } (A) \text{ AND } (B) \text{ THEN } (A').$$

There is more. It may happen . . . probably will . . . that the numbers of B molecules in the cell fluctuate. They might rise and fall with the energy level of the cell, or they could be most abundant when the cell divides. How will this affect you as an enzyme? Clearly, your production of A' will rise and fall as well—low B, low A' production, and so on. In biochemical terms, substance B causes a concerted change in your structure. This is allostery in action: a regulatory molecule binds

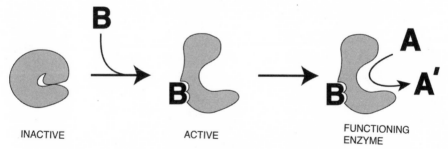

INACTIVE   ACTIVE   FUNCTIONING
                                                       ENZYME

FIGURE 3.3. An enzyme switch. A protein changes shape when it binds small molecule B. In its new shape, right, the protein becomes a functioning enzyme able to catalyze a chemical change (A to A').

to an enzyme (you) and causes a change in shape (switch to an active form).

What you are doing as an enzyme is analogous to what a transistor in an electronic circuit does (or a vacuum tube in one of Grey Walter's tortoises). The electrical current through the device can be thought of as the pipeline conversion by the enzyme. The controlling voltage, typically applied to the base terminal of the transistor, is like the small molecule that binds to the enzyme and regulates its activity. Small fluctuations in the concentration of B control the rate of A' production. And the quantitative relationship between the two need not be simple. Depending on the details of protein structure, the chemical output of the enzyme may be a highly amplified version of its input, as it often is for a transistor.

So you have a transistor. Or rather, stepping back, you have thousands of different transistors, since there are many different kinds of proteins in a cell. Each can perform a logical operation in which it, in effect, reads in the concentration of one small molecule and reads out another. What you need to do now is link these separate units, making a circuit. And the way to do this is simple, at least in principle. All you have to do is to share molecules. To link a series of enzymes into a chain, just make the product of one enzyme the raw material of the next. If enzyme 1 makes substance 1, enzyme 2 uses substance 1 to make

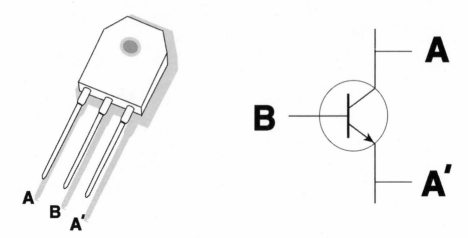

FIGURE 3.4. A transistor. A typical transistor has three electrical connections. Current passing between connections *A* and *A'* is modulated by voltage applied to connection *B*. A small change in *B* can result in a large change in A', as in a regulated enzyme. (Compare to the enzyme in figure 3.3.)

substance 2, and so on, you have in effect a production line. Provided you have starting material in sufficient quantity, these enzymes will convert it in a steady stream to a final product. Note that although I use the terms *production line* and *stream,* the enzymes do not need to be arranged in linear fashion. They may be bouncing about through heat energy, anywhere in the cell and in any orientation. The combination of diffusion and recognition through weak bonds will ensure that they work in the correct order.

To avoid becoming buried under symbols at this point let me return to the real world. What I have been describing is nothing less than the basis of biochemical reactions. Thousands of small molecules in cells—sugars, amino acids, lipids, nucleotides, and others—are chemically modified by enzymes. The small molecules are produced from the food we eat. In the process of digestion, food is broken down to smaller compounds with the generation of energy, much of it appearing in the form of ATP. The molecules we need for growth and repair are synthesized in other reactions, driven by energy. The hundreds of enzymes performing these chemical transformations are linked by diffusion and molecular recognition into large networks. They can be seen as metabolic maps adorning the walls of biochemistry laboratories, dense diagrams festooned with arrows, straight, curved, or branched.

Within this maze of cell reactions individual enzymes act like pipelines, converting one substance to another. Taken as a whole, the system is reminiscent of the plumbing that carries water in a large building. Or you might think in terms of electrical wiring. These analogies are useful—but you have to be careful. Remember that the arrows and positions on this map are illusory, drawn for your convenience because we humans like to think in spatial terms. In reality the molecules are jumbled together in the cell. There is not a single medium comparable to water or electric current that moves throughout the network. Biochemists like to talk of the flow or flux of molecules, but this is jargon. Molecules do not actually stream from one location to another: the terms actually refer to chemical changes, so molecules farther apart on the map are less similar in molecular terms. This metaphor between chemical space and physical space is crucial for the notion of biological computation, but it takes some getting used to.

The different enzymes in a cell are linked functionally because they share a common pool of molecules. Substrates, products, and regulators all come from the same set of chemicals. As one increases in concentration another falls. A small molecule such as an amino acid might be made by one enzyme and consumed by another. Another small molecule (such as the cyclic AMP described in Chapter 6) might be made by one enzyme but act as a regulator for many others. There are also global controlling elements, or substances that diffuse through the cell and control large numbers of different reactions. A good example is provided by calcium ions that are continually pumped out of the cytoplasm but flood into the cell in urgent circumstances, such as when a muscle contracts, or an egg is fertilized by a sperm. If you tried to represent all these reactions and connections in a diagram, the result would be complicated indeed, an impenetrable thicket of lines.

It would also be difficult to analyze mathematically. Large numbers of enzymes work together in ways that are hard to predict. Some aspects of cell chemistry have been studied for decades and are relatively well understood. The web of reactions that interconvert sugars and amino acids is a case in point: computer models and theories of metabolism are well advanced and relatively accurate. But even this is not yet an exact science. You would find it impossible to predict exactly what will happen if you increase the concentration of this enzyme, or change that feedback inhibition. Other parts of the network, such as the cascades of signals that control cell migration, say, or responses to hormones, are far less well developed. In these active areas of contemporary research investigators are still searching for key components and looking for crucial binding partners.

Fortunately, I do not need to know every detail of every last molecule in the cell for my argument. My aim is more modest—to understand in a general way how systems of proteins convey and process information.

FOUR

Protein Signals

Amoeba proteus is not one of those darling organisms like mice, fruit flies, and *E. coli* that have been chosen for in-depth analysis by molecular biologists. We have only a limited amount of information about its genes and proteins and mechanisms of motility. Giant amoebae have many special features such as streaming movements of their cytoplasm and a thick gel-like coating on their surface that are still incompletely explored. Moreover, even if we were fortunate in this respect and had a complete list of all the parts of an amoeba, we would still not be able to explain in detail how it works. There are many aspects of systems of molecules we still do not understand.

But if we stand back a little and ask about the broad features of the control systems and basic mechanisms of organisms in general, we might make better progress. Any free-living, motile animal must be subject to certain constraints, whatever its size. So if we examine the broad features of how a mouse, say, controls its behavior (in some respects better understood), then they could guide our inquiry into a migrating amoeba.

It's clear, for example, that to be independently motile, any organism must have sensory detectors. These have to be exposed to the outside world—for the mouse they are its eyes, ears, nose, and skin—and respond selectively to relevant features of the environment. We also know that sensory detectors generate signals relating to the size and type of the stimulus and convey them into the organism. In a mouse, the latter corresponds to the nerves extending from the sense organs to the brain or spinal cord.

Once signals are internalized, then they have to be processed. Organisms are inundated by stimuli patterned in space and time. They have to make sense of this sensory flood before they can do anything. One medley of smells combined with a tactile stimulus of the right kind might indicate the presence of a potential mate, requiring one course of action. A bright light plus elevated temperature might signal danger and trigger another response. To evaluate these different sensations—to compare them, give them different priorities—the organism has to perform computations.

What sort of computations? Perhaps the organism needs to know the cumulative effect of two or more stimuli and therefore has to perform an addition. Or it might want to know how much stronger one is than another, equivalent to a subtraction. Or the animal might amplify one kind of signal according to the level of a stimulus, as though turning the volume dial of a radio. Or it might be crucial to know the times of arrival of specific signals—such as the flap of a wing and an approaching shadow. If a sequence of periodic impulses arrives from an outside source, the organism might benefit from knowing their frequency. Or if a scent rises and falls with time, perhaps knowing how fast this happens will help—indeed, bacteria use something like this in their search for distant sources of food. The possibilities are endless. For a mouse these computations take place in the central nervous system. Once a decision has been made, signals must be sent to the appropriate organs for action. In a mouse, electrical signals travel along nerves to muscles and cause them to contract.

I would like to convince you that systems of protein molecules can perform all of these tasks. They provide a single cell with the equivalent of sensory organs, nerves, and muscles. Made as chains of amino acids of unique sequence, proteins fold into a small number of stable shapes and can flip from one to the other depending on conditions. They interact with other molecules in the cell through the universal processes of diffusion and formation of weak chemical bonds. These features are ubiquitous, fundamental, and essential. But evolution has elaborated this basic machinery, has provided special tricks that help transmit signals and process information.

The detection of environmental signals, for example, calls for proteins of a special sort, known as receptors. The skin of a cell is its membrane, a thin oily layer made of lipids and proteins that encloses the watery cytoplasm and defines its boundary. Any protein acting as a sensory detector must therefore be at least partly immersed in the membrane. Complete immersion is difficult. Individual proteins are so large that portions bulge out into the surrounding water, like globular icebergs floating in a lipid sea. Receptors are embedded in the cell membrane with one portion exposed to the outside of the cell and another exposed to the cytoplasm on its inside. They are ideally suited to convey information from one to the other.

Cells have hundreds of different kinds of receptors. Each is specialized to detect a particular substance—perhaps a protein, or a peptide, an amino acid, a sugar, a steroid, a fatty acid, or even a dissolved gas. The physical basis of recognition by a receptor is selective lock-and-key binding. Each receptor has a binding site that it presents to the outside world. If a suitable molecule is carried by diffusion to this open mouth, it fits snugly through the formation of many weak bonds. A cascade of internal changes in the receptor then ensues, causing the receptor to switch its shape. This shape change flits across the cell membrane—usually in much less than a millisecond. The cell has now officially registered that it has encountered a molecule of a particular type.

The next step is to convey this knowledge to other locations. I already mentioned one mode of transmission: changes in a cascade of enzyme reactions. The first enzyme makes a small molecule that serves as the starting material for the second enzyme, the second makes starting material for the third, and so on. Evidently, if enzyme 1 suddenly increases its activity (because it is attached to a receptor that has just received a signal) then that change will pass to the other enzymes and the molecules they influence. It's a perfectly feasible way to propagate a message . . . though not a very efficient one.

A faster and more controlled method of relaying signals, widely employed by cells, is via the chemical modification of proteins. Instead of using the binding of a small molecule to trigger a protein switch, cells employ enzymes that attach groups of atoms to a protein through strong bonds. Remember that the precise way a chain of amino acids folds is extremely sensitive to the interactions between its atoms. Small perturbations of any kind can change these interactions and might cause—if these have been selected by evolution—an abrupt change in form. Thus there are proteins that sense the voltage drop across a membrane—vital for the electrical signals produced by nerve cells. There are proteins that respond to local physical forces, sensing, for example, the stretching of a cell membrane as a cell imbibes water. So it's no surprise that if you attach a chemical group to even a single amino acid, then this can cause the protein to switch from one shape to another.

The incredible variety of enzymes in a cell includes many that work on other proteins. Typically such an enzyme might attach a small chemical group to its protein target, such as a methyl group (carbon and three hydrogen atoms, written as CH_3) to a particular amino acid. Or it might add an acetyl group (derived from acetic acid), or a sugar, or a lipid. These groups are physically small compared to the whole protein, no more than a button on a blouse. But when added to the right spot—and we can leave it to evolution to find this spot—the new group can trigger a seismic change. In other words, it causes a cascade of atoms to shift in position, so that the protein molecule flips to a different shape. Exactly the same arguments I made in the previous chapter for the binding and unbinding of small molecules can be applied to the attachment of a new group. Once again, it provides a functional link between two otherwise unrelated processes in a cell.

There are differences, of course. Chemical changes usually last longer than simple binding events. If the cell needs to reverse the shape change quickly, then it must adopt a second enzyme to do so. Moreover, the chemical modification of a protein requires an input of energy—

participation of an activated, "high-energy" intermediate that drives the required reaction to completion.

Although cells modify their proteins in many different ways, the simple addition of a phosphate group is by far the most common. In fact, almost half of all of protein molecules in a cell can carry one or more phosphate groups, an amazingly high proportion. Enzymes that add phosphates to other molecules, or kinases, are correspondingly numerous. Human DNA encodes genes for several thousand different kinases, each with its own molecular target.

Strong chemical bonds are long-lasting. Their strength ensures that they resist the disruptive effects of thermal energy for long periods of time. So they could be relatively permanent. For example, the addition of phosphate groups to a protein in nerve cells encodes long-lived memories in the brain. But from the standpoint of propagating messages, we need protein molecules that can switch on and off, preferably very rapidly. And for this reason most kinases in a cell are paired in the cell to a second enzyme that reverses their effect—a phosphatase that removes the phosphate attached to the protein.

In a typical signaling pathway, proteins are continually being modified and demodified. Kinases and phosphatases work ceaselessly like ants in a nest, adding phosphate groups to proteins and removing them again. It seems a pointless exercise, especially when you consider that each cycle of addition and removal costs the cell one molecule of ATP—one unit of precious energy. Indeed, cyclic reactions of this kind were originally labeled "futile." But the adjective is misleading. The addition of phosphate groups to proteins is the single most common reaction in cells and underpins a large proportion of the computations they perform. Far from being futile, this cyclic reaction provides the cell with an essential resource: a flexible and rapidly tunable device.

A good way to think of the operation of a cyclic reaction is in terms of water. Picture a tank that is being filled with water from a tap and simultaneously leaking water from a spigot in its bottom. As the water in the tank rises, the rate of loss through the spigot will increase, since it depends on hydrostatic pressure. Eventually the rate of entry and the

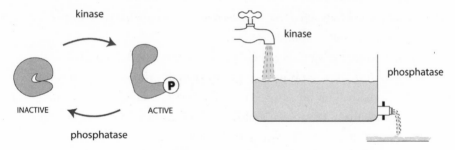

FIGURE 4.1. A cyclic reaction. On the left, a reaction is shown in which a phosphate group (P) is continually being added to a protein and then removed again. Two enzymes, a kinase and phosphatase, separately control phosphate addition and removal. On the right, the concentration of the protein with phosphate is represented by the level of water in a tank and the two enzymes as input and output flows.

rate of loss will exactly balance, and the water level will reach a steady state, technically known as a dynamic equilibrium.

Now consider what will happen if you open the tap further. Water now enters the tank faster than it can leave. The level in the tank rises to a new level until the exit from spigot again balances the tap. Or, if you wish, you can achieve the same result by shutting the spigot at the bottom instead of opening the tap at the top: the final level depends on a balance between the two. Another insight gained by this analogy concerns how quickly the water level can respond. If both tap and spigot are passing a slow trickle of water, then changes in the tank will take a long time to build up. But if tap and spigot are spouting furiously, any change in one or the other will produce a vigorous rise or fall in water level.

The level of water in this analogy represents the average number of phosphate groups in a population of molecules of a certain type. The tap represents a kinase that adds phosphate groups, and the spigot a phosphatase that removes them. An ability to control the tap and the spigot is realistic because kinases and phosphatases are without exception regulated. They are themselves switches controlled by other molecules linked to other pathways.

The cyclic reaction is a logical device. Once again we can liken it to an electronic switch, in this instance one made from three transistor equivalents. The central transistor provides the output of the device (the protein having a phosphate added and removed). The other two transistors are the two enzymes that add and remove phosphate groups, respectively. These additional controls add several advantages. First of all, they allow the entire cyclic process to be driven by ATP. In our analogy gravity caused water to fall into the tank and out at the bottom. The equivalent driving force in a cyclic reaction is chemical energy. Phosphate groups start off in a high energy state in ATP, they pass to a slightly lower energy state attached to a protein, and they finish at the lowest energy state of all as simple phosphate ions in solution. In this way the enzymes of a cyclic reaction plug into the power supply of the cell, using chemical energy released from food.

So our computational protein can be modified as rapidly as the cell requires. In chemical terms it can spin like a wheel, with phosphate groups being attached and detached many times a second. If the cell really needs to change the concentration of the modified protein very quickly, it can. All it has to do is to switch on or shut off the phosphate-adding reaction and the concentration will fall precipitously—at the speed of the spinning cycle. There is no buildup of products or depletion of substrates to slow down the process, as there would be in a linear chain of enzyme reactions.

I mentioned above that the enzymes that add and remove phosphate groups are themselves regulated—for example, by binding to small molecules. Earl Stadtman and his colleagues examined the capabilities of this common strategy in 1977. They studied the cyclic conversion of a specific enzyme between two forms, one with and one without phosphate. They demonstrated, first by theoretical analysis and later by actual wet biochemistry, that the input/output responses of such cycles could be modulated over a wide range. Many similar studies have subsequently confirmed and extended these results. The time course of response of these enzymes—how they rise and fall—can be of almost any shape. In particular, they can if necessary be very sharp, like digital switches.

Small chemical modifiers such as phosphate have one further advantage: they can provide a protein with many inputs and outputs. Consider the enzyme *glycogen synthase*, a key element in the body's energy balance. As its name implies, this enzyme catalyzes formation of glycogen, a polymer of glucose used as energy storage by cells. And because energy balance is highly important to the organism, this enzyme is subject to elaborate controls. When we have just eaten a large dinner and collapsed into an armchair, glycogen synthase in the liver works at high efficiency, building small granules of glycogen. Conversely, if we've been performing hard physical work all day and have skipped lunch, glycogen synthase is turned off while other enzymes break down glycogen, releasing sugar for use in the brain, in muscles, and elsewhere. We understand the molecular basis of this switch in considerable detail, and it illustrates both kinds of regulation we have met. First, glycogen synthase is activated when it binds to a small molecule—a sugar phosphate that rises in concentration when cells have abundant glucose. Second, it is inhibited by the attachment of phosphate groups to specific amino acids.

But unlike the idealized enzyme described before, this real enzyme has at least nine sites for phosphate groups! At least four different en-

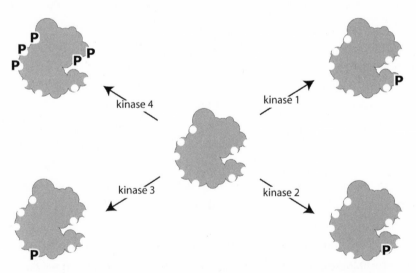

FIGURE 4.2. An enzyme with many controls. The enzyme glycogen synthase has nine sites on its surface that can receive a phosphate group (P). Four enzymes (kinases) add phosphate groups to these sites, at rates that fluctuate with conditions in the cell. Consequently, glycogen synthase exists in many forms, each with a particular set of phosphate groups.

zymes (kinases 1, 2, 3, and 4 in figure 4.2) catalyze these additions. A further complication is that some of these sites must be occupied before phosphate can be added to the others—that is, regulation is hierarchical.

It's almost as though glycogen synthase is a strange musical instrument with ornate controls on its surface, like stops and valves. One valve is depressed when the enzyme meets and binds a small molecule; other valves are depressed as phosphate groups are added or removed. Some of the chemical forms thus created have different activity; others may have different spatial locations in the cell, or perhaps interact differently in the social milieu of the cytoplasm. It is possible that every one of the dozens of different possible "notes" played on this protein trumpet is essentially unique.

There are many other kinds of modifications of protein structure. They constitute a feature unique to living systems and currently have no electronic equivalent.

One of the foundations of the field of artificial intelligence is a paper published in 1943 by a neuropsychiatrist, Warren McCulloch, and a mathematician, Walter Pitts. They demonstrated in this paper that small numbers of idealized nerve cells, linked into circuits, could perform essentially any logical operation one wished no matter how complicated. The authors declared: "Because of the 'all-or-none' character of nervous activity, neural events and the relations between them can be treated by means of propositional logic." Although it was recognized from the start that these devices (or McCulloch-Pitts neurons) were artificial and highly simplified, there was no reason, it was argued, why real nerve cells in the brain should not be capable of performing equally clever operations. Few today would disagree with this thesis.

As with neurons so with proteins. Because protein molecules can perform logical operations, we can, if we wish, represent them as a form of computational element similar to a McCulloch-Pitts neuron. Protein molecules provide the cell with a tool kit of components to build virtually any circuit, putting aside for the moment the question of *why* it would want to do this.

Imagine, for example, a protein with two possible phosphate sites (like the glycogen synthase trumpet but with just two keys). Even such a simple construct could perform logical operations. The protein might be active only when both sites have a phosphate group attached, inactive otherwise. Or it could be active if just site A, but not site B, was occupied. Or it is conceivable that either site A or site B would work . . . and so on.

The operation of this notional enzyme would then conform to a system of mathematical logic invented by the English mathematician George Boole in the mid-nineteenth century. Boolean algebra provides a systematic account of the possible logical interactions of two inputs using operators with names such as AND, OR, and NOT. The situation in which both phosphate groups have to be in place corresponds to an AND switch; if either of the two phosphates will do the job, it is an OR switch; and so on. Boolean algebra started life as an abstruse form of pure mathematics but became unexpectedly relevant to practical matters with the rise of computers. The notion that biological molecules such as enzymes might also act as Boolean switches has been extensively explored. As a proof-of-principle, investigators have made Boolean switches out of biological molecules in the laboratory, using enzymes, genes, or small sets of genes.

Boolean switches rarely exist by themselves, whether in a computer or a living cell. They are almost always connected into larger modules with more complicated properties. Connections are made when the output of one switch acts as an input to a second. For example, if one enzyme makes glucose and a second enzyme converts glucose to another sugar, then the two enzymes are functionally connected. Another way to make a connection is if the output of one enzyme is a regulator of another. The enzyme that we discussed in Chapter 3 that catalyzes the reaction $A \rightarrow A'$ and is regulated by B, acts as a connection between three substances, A, A', and B.

And these are simple enzymes. The example of glycogen synthase above shows us that a single protein can act as a point of convergence for many inputs. Conversely, a kinase that adds a phosphate group to

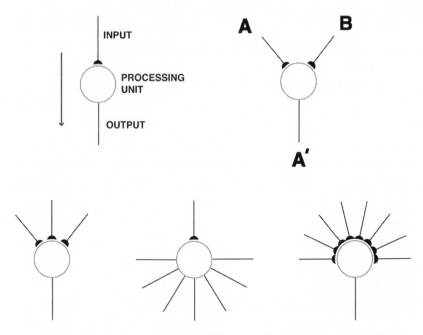

FIGURE 4.3. Enzymes as computational elements. A single processing unit in a computational network performs a logical operation on its input and produces an output (upper left). Similarly, an enzyme or other protein in a cell (upper right) typically binds to one or more other molecules (in this case A and B) and modulates its function (production of A') depending on their different concentrations. The lower half of the figure displays arrangements of inputs and outputs encountered in real enzymes.

many different proteins (as most do) allows a single input stimulus to fan out in many directions. Given a toy box of these tiny processing units we could link them in almost any way we wished.

Sets of linked proteins work together to perform simple tasks. Three enzymes in a marine bacterium make an oscillator, adding or removing phosphate groups according to the time of day. A cascade of kinases in yeast filters out rapid fluctuations in the osmotic strength of the surroundings while allowing the cell to respond meaningfully to slower changes. Enzymes degrading sugars produce a slow-on-fast-off dynamic reminiscent of the closing of an elevator door (which stops the second a foot is put in the way). There are hundreds of other possible circuit motifs. Rather as the words of a language are put together in different combinations to build sentences and paragraphs, so circuit motifs are used to construct large biological networks.

Using these elementary circuits, we can in principle design any circuit we wish: link them in biochemical space. This is the basis for the large networks of metabolic reactions familiar to biochemists. It is also the way that such substances as hormones, acting on the outside of a cell, cause signals to spread into its cytoplasm and nucleus. Of course, designing such a circuit from scratch is not a trivial matter: your reactions have to have the right specificities and operate at a suitable speed before the circuit as a whole works as required. One of the easiest ways to achieve this is by trial and error. Set up the connections on a computer and then change the rates of the different reactions. After each change, run the network and see how close it comes to doing what you want, a procedure with some similarities to evolution.

I'm not suggesting that living cells work out abstruse logical or mathematical functions for their own sake. A liver cell does not want to calculate logarithms to base 10 or find the cube root of a large number. Cells have a different agenda. Despite the analogies we have noted, there are also many differences in both the "hardware" and "software" that cells use compared with electronic devices.

One feature of electronic computers that, at first sight, seems to make them different from living systems is that they work in digital mode. Televisions, mobile phones, and laptop computers carry signals in long sequences of ones and zeroes. That is, they channel electrons in streams that alternate, intermittently and in precise sequences, between a basal level represented by 0, and a maximum level represented by 1. Similarly, the hardware components used in such devices are often designed to work in an all-or-none fashion. A typical transistor, for example, works by combining voltages in as close as possible to an idealized logical, or Boolean, fashion. They also act very quickly and a typical electronic circuit carries many millions of bits (zeroes and ones) every second.

Living organisms, by contrast, seem to be incorrigibly analog. In other words, they do not change in stepwise increments, in ones and zeroes, but as a continuous variable over a wide range. If you measure virtually any feature of individuals in a population of mice or men, such as color, weight, or size, you will find a smooth distribution—often with a single broad peak. A culture of skin cells under a microscope displays a jumble of organelles and structures that vary in number, size, shape, and position in an apparently capricious manner—nothing digital there. Similarly, in the normal day-to-day running of our bodies, parameters such as temperature, metabolic rate, and blood sugar rise and fall in a continuous analog fashion.

But a moment's reflection reveals that the difference between living and nonliving computers is not clear-cut. Most electronic devices also display analog features. Heat, light, and sound are all of an analog nature, so that thermostats, cameras, and microphones have to read inputs that vary continuously over a range of values. They typically convert these inputs into streams of zeroes and ones in order to perform digital calculations—and then convert them back to an analog format for the benefit of our human senses. Nor is it even necessary for calculations to be performed in digital mode. There are analog computers that convert variables into electrical quantities such as current and voltage. Rather than providing a discrete yes or no answer, analog computers indicate how much, typically producing an electrical current that has a distinct value between two extreme limits. The current might rise and fall so as to represent the flight path of an airplane or wind speed in a hurricane.

Just as electronic devices are not necessarily digital, so biological systems are not necessarily analog. We see this at the level of the genes. Molecules of the hereditary material DNA are made as a linear chain, or polymer, of smaller chemicals linked by strong bonds. The repeating subunits are not amino acids, as in protein, but nucleotide bases: four types, designated by the symbols A, C, T, and G. Their sequence in the chainlike molecule (for chemical reasons DNA does not fold into compact structures like protein) encodes the genetic information of the organism. From a computational standpoint, therefore, we could represent each base as a two-digit binary number and not lose any information. For example, we could substitute 00 for every A, 01 in place of C, 10 for T, and 11 for G. A series of bases in DNA such as

AATGACTAGACCTGGTAA . . .

would then be replaced by

0000101100011000110001011011111100000 . . .

a string that is indubitably digital in format. Individual genes encoded in DNA are often switched fully on or off by external factors (usually proteins), like binary switches. The complicated logic of developmental processes (often referred to as developmental programs) arises from circuits built of genes and the proteins that control them.

Even protein molecules have digital features. Allosteric proteins adopt one of a small number of discrete conformations and flip between these states in a rapid all-or-none fashion. They make switches with very sharp transitions, such as those that carry signals in the brain and nervous system. Channels and pumps carrying ions into and out of nerve cells create electrical responses that are characteristically abrupt and all-or-none. They can be detected and measured by positioning a special electrode close to the molecule in the membrane. If the results are displayed on an oscilloscope, a device that records changes of voltage with time, they show a series of square-wave pulses—abrupt rises and falls in voltage that are as close to a digital output as you could wish.

So both living systems and electronic devices make use of both digital and analog technology. Each format has its advantages. Analog circuits can be very rapid and often require fewer components. Digital circuits have the advantage of greater reliability and are better for the transmission of information in noisy situations.

In reality, it seems that many circuits in living cells employ a strategy somewhere between the two extremes, sometimes referred to as fuzzy logic. Household devices such as vacuum cleaners, washing machines, and refrigerators often incorporate fuzzy logic circuits. A room heater might be controlled by ambient temperature, humidity, time of day, and user preference. These different parameters, sent to the heater control circuit, would be put together in a judicious manner—almost like a person deciding whether to turn a control knob up or down. For example, the room might be cold but not humid; it might be after 8 P.M., when electricity is charged at off-peak rates; I might prefer a higher room temperature at the end of the day. Each of these factors can be given a numerical value and combined mathematically in the heater circuit in determining its actual output.

We can gain some insight into the switches actually employed by cells by considering how they use nitrogen. This is a crucial element for all living systems and many enzymes are engaged in its distribution. The enzyme glutamine synthase sits in a particularly influential position, like the signal box controlling a network of railway lines. It incorporates ammonia into the amino acid glutamine, which then serves as a readily accessible form of nitrogen for many other essential compounds, including the subunits of protein and DNA.

The enzyme itself is an assembly of twelve subunits in a double ring structure and is regulated by all the tricks mentioned in this chapter. Small molecules bind to sites on the enzyme, causing activation or inhibition. It is also chemically modified—in this case the groups added to the enzyme are not the familiar phosphates, but they affect catalytic activity just the same. Once again the precise activity of the enzyme depends on the modifications made. A constellation of influences from other pathways conveys information on the current state of supply and demand of nitrogen.

This situation leads to a complicated picture, reminiscent of a device controlled by fuzzy logic. Each of the inputs to glutamine synthase makes a contribution that is weighted according to its importance. This enzyme can recognize different combinations and change its activity accordingly. Its performance is therefore far richer than that of a single digital transistor.

Returning to our crawling amoeba, I've shown how special proteins, receptors, inserted into the membrane serve as sensory detectors. Another larger set of proteins in the cytoplasm integrates messages coming from these receptors, processes them, and conveys them to every part of the cell. Now, at the last stage of the pathway, the signals must take effect—control the movements of the organism.

Most animal cells move by means of a system of protein filaments known as the cytoskeleton. Filaments extend throughout the cytoplasm, attaching here and there to the membrane, the nucleus, and other structures. The cytoskeleton provides not only structural support (as implied by the term *skeleton*) but also the driving force for movement, acting in this sense more like a muscle. Indeed, the muscles in our arms and legs are made of huge arrays of protein filaments that generate contractions by sliding over each other. Animal cells that crawl over surfaces, such as amoebae and the white blood cells that wander through our bodies looking for invading bacteria or dead or dying cells, employ motile machinery based on a protein called actin. This forms long thin filaments that assemble into a variety of networks, gels, and bundles through their interaction with a host of other special proteins. Another class of protein filaments called microtubules provides tracks for the active transport of molecules and organelles within the cell. Microtubules form the distinctive spindle seen in dividing cells that carry newly formed chromosomes into the daughter cells.

The details are enormously complicated. Cascades of signaling reactions spreading from receptors in the membrane eventually reach the cytoskeleton and change its state and performance. Interactions between protein molecules promote the growth of actin filaments. The lengthening actin filaments push the leading margin of the cell forward (and pseudopodia extend). Elsewhere in the cell, other signals trigger interactions between actin and motor proteins such as myosin. The latter has the capacity to move along actin filaments using the ubiquitous ATP as an energy source. Myosin movements pull on actin filaments, causing them to bunch up and contract—this is basically how the cell retracts its trailing end. Coordination of these multifarious activities in different parts of the cell (a major problem for a cell as large as *Amoeba proteus*) results in integrated movement.

How then—returning to the question I asked at the start of this chapter—do molecules inside a cell work like a nervous system? The answer is that they detect environmental stimuli and codify their modality and strength. They relay information about these signals to different locations of the cell and combine them with other signals. Molecular systems process signals so as to produce a logical output. Interacting with other systems of proteins, specialized to produce mechanical rather than chemical effects, the output signals produce physical movements. The cell moves in an informed if not intelligent way.

It is a very different kind of computational machine from those we are familiar with, and it takes some getting used to. There are parallels between the action of individual protein molecules and that of the components of an electronic circuit. But there are also huge differences. A protein molecule is perhaps a thousand times smaller in linear dimension than a silicon transistor. As though in compensation, however, it is usually present in thousands of individual copies, scattered in different locations in a single cell. Many molecules, especially the small ones, are in continual violent motion owing to thermal energy. How can you connect circuits under these conditions? Where are the wires?

The answer is that the wire-equivalents are molecules that diffuse and interact. Typically, a molecule made by one enzyme diffuses a short distance to the second enzyme. En route, it makes fleeting contact with thousands of other molecules. But it disregards these encounters and comes to rest only when it reaches the second enzyme. Only this protein has a binding site that matches the contours of the diffusing molecule so tightly that it binds and sticks. The first enzyme has now influenced the second, not through the passage of electrons in a copper wire but by the physical movement of a molecule.

So there are no wires. In fact, the term *biochemical circuits* is flawed in several respects, a product, no doubt, of our propensity to attach spatial metaphors to processes of all kinds. In reality, a signal traveling through a cell is a change in the numbers of specific molecules at particular locations. Signals move from one place to another by diffusion and the influence of enzyme catalysis. It sounds like a clumsy and haphazard process to us. But in the world of atoms and molecules it can be astonishingly rapid and efficient. Let's hope it is anyway, since our thoughts and actions depend on this very mechanism! Synapses that link one

nerve cell to another or connect a nerve cell to a muscle depend on precisely this same sequence of diffusion and binding. Molecules (neurotransmitters) released by signaling activity of one cell diffuse over a tiny
gap (about the length of one hundred atoms put end to end) to the second cell, where they bind and activate a new signal. It all happens in
much less than a millisecond.

We might also compare how connections are made. In an electronic
device such as a television, functional units of different sizes are connected by a highly logical arrangement of electrically conducting links. In
comparison, the wiring of a cell rests ultimately on the atomic structures
of its large molecules. Their structures specify the molecules they will interact with and what effect this will have—whether, for example, a protein
will be modified by addition of a phosphate, and where (at which amino
acid).

Ultimately, the wiring diagram—or, more accurately, the recipe for
making the connections—is written in the sequences of base pairs in
DNA, since these specify the three-dimensional shape and surface chemistry of proteins. But the actual formation of connections themselves is
left to the haphazard process of thermal motion. It's as though a collection of electrical components—resistors, transistors, capacitors—had
been labeled with individual numbers, put into a bag, shaken so that their
positions were randomized, and then connected according to their numbers. This would be a nonsensical task for an electronics engineer: for one
thing, there would not be enough space for all the wires. But for a living
cell full of freely diffusing molecules, it is a highly practical solution.

Were we able to enter the cell like miniature voyagers and see its molecules as lights of different colors, we would have an overwhelming kalei-
doscopic experience. Pulsing, flashing lights, like fireworks moving in
intricate patterns, would indicate, by their color and frequency, the local
concentrations of molecules in the cell and hence its current physiological state. If we knew how to read this code, it would tell us what the cell
had eaten, what molecules it was making, how close it was to dividing.
Perhaps it would even tell us what the cell knew of its surroundings, and
of itself.

Cell Wiring

The difficulty with general principles in biology is that they are buried in a mountain of details. The theory of evolution came to Charles Darwin only after years of interest in natural history and an extended sea voyage that immersed him in a riot of natural variation. Biology's most powerful generalization was distilled from a myriad of special cases, like alcohol from a mass of fermenting grain. Even today, although few scientists question the validity of Darwin's theory, the way it operates can be subtle and hard to define.

You can say the same about cellular computations. Although I've described in an abstract way how protein molecules make connections and perform logical transformations, it's difficult to grasp how these work in practice. Try mapping all the protein-protein interactions in even a small region of the cell, and you will create a mass of densely interwoven lines impossible to unpack. So, as for evolution, the best way to understand how protein computation works in a cell is to follow a single thread—choose an exemplar that reveals basic mechanisms with particular clarity. And one of the best places to look, I'd argue, is the way that bacteria forage for food. All of the principal components of this signal pathway are known, from input to output. By going through it step by step, I will show how protein molecules underpin a simple form of cell behavior. Lessons learned from this circuit can be applied to higher organisms, including humans.

Chemotaxis in the bacterium *Escherichia coli*—its ability to swim toward distant sources of food or away from noxious stimuli—depends on a small set of proteins. The pathway begins, logically, with receptors: molecules embedded in the membrane that look at the outside world. Because of binding sites on their surface, receptors recognize certain small molecules. Substances such as amino acids and sugars that indicate the presence of food are treated as attractants, whereas potentially harmful substances such as nickel ions and acids are treated as repellents. Although only four kinds of receptor survey the surroundings, each of these is multifunctional so that as a whole they enable the bug to detect more than fifty attractants and repellents. This is its chemical universe. A typical bacterium has about ten thousand receptor molecules in all, most of them clustered at one end of the cell.

It also has flagella, typically four to six distributed randomly around the cell, together with the rotary motors that turn them. As we saw in Chapter 1, these motors cause the bug to tumble by switching their direction of rotation and hence steering it toward distant sources of food. Somehow these two processes—detection of signaling molecules at one end of the pathway and rotation of motors at the other—are coupled. This "somehow" is crucial—the wiring of the response.

The driving force for the pathway is a cyclic reaction that first adds a phosphate group to a protein and then removes it. A protein attached to the inside of the receptors actually performs this cycle solo, serving as both kinase and phosphatase. It transfers a phosphate group from ATP to one of its amino acids and then, after a while, releases it again in an unending cycle. The receptors speed the reaction several hundredfold, depending on their state. Driven by the energy available in ATP, the kinase cycle both provides an amplification of the incoming signal and acts as a switch to turn the response on and off crisply.

The kinase can transfer its phosphate to proteins downstream in the signal pathway. One of these, CheY (Che for chemotaxis, pronounced "key"), provides the crucial link with the motors. Shuttling back and forth by diffusion, the phosphate-carrying form of this small protein, CheYp, regulates which way the motors turn. High concentrations of CheYp cause the bug to tumble (clockwise rotation of motors), whereas low concentrations suppress tumbling and encourage the bug to swim smoothly (counterclockwise rotation).

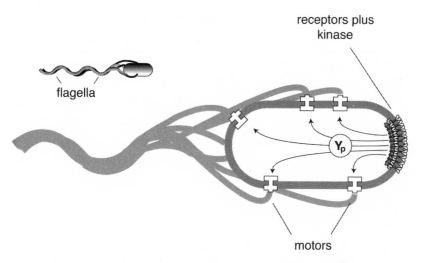

FIGURE 5.1. Signals that control bacterial swimming. Receptor proteins inserted in the cell membrane detect chemicals of interest in the surrounding fluid. A kinase associated with the receptors adds a phosphate to a protein, CheY, at a rate that depends on external stimuli. The product CheYp (Yp) then diffuses through the cytoplasm to the flagellar motors, where it influences their rotation.

The first lesson the chemotaxis pathway has for us is that its protein molecules are not all floating freely. Although they are perfectly capable of diffusing in the cytoplasm and membrane and of finding their partners, however distant, they are for the most part immobilized. At the detection end of the pathway, tens of thousands of protein molecules, including the receptors and the kinase, congregate into a flat irregular disc. The long thin receptors poke through the membrane so that they see outside the cell, whereas other proteins are stacked on the membrane's inner face, immersed in the cytoplasm. Weak bonds hold adjacent proteins together with a precise orientation and strength so that the structure has a characteristic architecture. Evidently evolution has gone to considerable trouble to make this tiny structure, but why?

The answer is that proteins gain extra functionality by working together. The whole protein complex has greater capabilities than the sum of the constituent parts. In the case of the receptor cluster, one benefit is sensitivity. Through its weak bonds the system is poised so that a change in just a few individual receptors can trigger a landslide of events involving a larger number of kinase molecules. Even the slightest whiff of food startles the bacterium into a positive response. The cluster also helps discriminate between different flavors and provides a framework for the memory of the bacterium.

Living cells are full of structures made from many protein molecules: protein complexes. The cytoplasm is not so much a thin watery bouillon soup as a thick fish chowder, with large and small lumps made mainly of protein. Some biochemical computations do occur in solution, it is true. But many more take place in a solid, condensed state, with chemical and shape changes propagating from one protein to another through direct physical contact. The receptor cluster in the bacterial membrane detects and processes signals from outside the cell: it not only amplifies the input signals but also analyzes their time course. Protein complexes in the synapses of the brain have a similar but more complicated processing ability. Elsewhere you can find protein complexes with moving parts that function like miniature motors. Some, driven by ATP, move from one location to another within the cell along protein tramlines. Others, such as the barrel-shaped structure that attaches a bacterial flagellum to the cell, are made to rotate by ions flowing across a membrane. Over the past few decades, protein machines of amazing variety and sophistication have been discovered in cells. They perform all of the basic functions of a cell, including the copying and repair of DNA molecules, the control of gene expression, and the manufacture and degradation of protein molecules.

Protein molecules in a large complex interact in highly specific ways. Not only do they occupy a particular location, with a defined set of neighbors, but their actions are also ordered in time. A series of reactions or shape changes propagates through the complex, each change depending on the previous one. This organized, solid-state property, resembling the actions of a piece of clockwork or part of an engine, has led to these assemblies also being described as protein machines. The description is particularly appropriate when the primary function of the machine is to produce movements, as with the molecular motors that carry small membrane-enclosed sacs (vesicles) and other cargoes along protein tracks from one location to another. Watch one of these under a microscope and you will see it cruise at a steady rate for a period, then stop and start again. But protein machines can also have a chemical function. A protein complex that moves along a chromosome making a copy of DNA, or one on the surface of a plant cell that spins out cellulose fibers, will go through just as many intricate internal changes of conformation.

The distinction between chemistry and mechanics is a human invention and not one that concerns a cell. At the atomic level, all movements entail a chemical change and all chemical changes create movements. The difference is one of degree rather than kind.

I can go further. Like individual microchips in an electronic circuit, protein complexes carry out sets of logical operations. Sometimes the result is a conspicuous physical movement—something you can watch under a microscope. At other times it is a chemical transformation that can be detected only by a sophisticated assay. But the underlying principle from the standpoint of this book is the same—they are all forms of computation. Although built from dumb molecules, protein complexes somehow operate at a higher level. They seem to have taken a step in the direction of life.

In point of fact, *E. coli* chemotaxis operates by means of *two* protein complexes. One is the membrane-bound cluster of receptors and associated proteins just discussed; the other is the motor. The bacterium typically has four to six motors disposed at random locations on the cell surface. Each is capable of spinning its attached flagellum several hundred times a second, and of reversing direction within a tenth of a second. Built from more than forty different proteins, the flagellar motor has a ring of static elements embedded in the membrane and a central rotor. It has other parts needed to control direction and to direct its own assembly (remarkably, the motor and flagellum continue to grow in length while rotating). Rotation is driven by an influx of hydrogen ions (protons) flowing back into the cell through eight protein channels, rather like water in a turbine.

Communication between the receptor complex and each of the motors is by diffusion. Molecules of CheY pick up a phosphate from the receptor complex at a rate that reflects conditions outside the cell. They then diffuse through the cytoplasm until by chance they encounter one of the flagella motors. There they stick: the more molecules of CheYp that bind, the more likely the motor is to spin clockwise and therefore generate a tumble.

Proteins change their shape at every step in this pathway. The receptors exist in at least two distinct states, stabilized by attractants and repellents. Because the receptors are intimately intertwined with proteins such as the kinase on the inner face of the membrane, their state influences that of their neighbors. Changes in shape propagate. The central kinase itself acts like a two-stroke motor, switching between two states as it creates active phosphate groups. Then, when a phosphate is transferred to CheY, the latter protein changes its shape and diffuses into the cytoplasm. Binding of CheYp to the motor initiates another cascade of shape changes, causing the motor to switch the direction of its rotation. These are just a few of the principal changes. The complete story would be complicated indeed, with molecules flipping between states with stroboscopic intensity in every part of the pathway, from receptors to motors.

Now I can give bacterial memory a molecular explanation. Bacteria store a running record of the attractants they encounter. This tells them whether things are better or worse: whether the quantity of food molecules in their vicinity is higher or lower than it was a few seconds ago. It's a pragmatic strategy: if conditions are improving, continue swimming; if not, tumble and try another direction.

You can investigate the origins of this running record in more detail by exposing bacteria to a step change in the concentration of an attractant. Now it is clear that what the bugs respond to is not the concentration of aspartate per se but its rate of change. A sudden rise or fall of aspartate creates a signal—changes the tumble frequency or the level of CheYp. But once aspartate has settled down to a steady concentration, the bug no longer responds. Biologists call this adaptation, but a mathematician examining the time course of response would call it differentiation. By measuring the rate of change in the signal, the receptor cluster has in effect performed calculus!

How does it do this? The answer, in a word, is methylation, the chemical addition of a methyl ($-CH_3$) group. This is another type of protein modification akin to the addition of a phosphate. Two enzymes in the receptor cluster continually add and remove methyl groups from the receptors. The methyl groups, rising and falling in number, control the signals sent from the receptors, working somewhat like the volume control on a radio.

Imagine that you are listening to a distant radio program that fluctuates in volume in an unpredictable fashion. Suppose that you are trying to keep the sound intensity constant by turning a volume control that is, unfortunately, stiff to turn. Sudden rises and falls are too fast for you: the sound either disappears or it blasts your eardrums. But let the signal settle down and you easily reach a balance. In similar fashion, the number of methyl groups on receptors rises and falls with the signals entering the cell. Methylation tends to cancel out the signal caused by the external stimulus. It lags behind the signals because the enzymes work relatively slowly, and this is why the response of the bug reflects the rate of change of external attractants.

The number of methyl groups carried by receptors—like the setting of the volume knob—is a measure of the concentration of attractant in the recent past. Each receptor has eight sites of possible methylation and can have anything from zero to eight methyl groups. Zero means the volume knob is turned fully down (a condition that might occur in response to repellents). Eight means a volume knob turned to its highest value (a huge, saturating dose of attractant). Most receptors most of the time will be somewhere between these two extremes. But the precise number of their methyl groups and their distribution between the different kinds of receptors—recall that the bug has four distinct species of receptors—reflects the recent environment sampled by the bacterium. This indeed is the memory I have referred to. It is a record that tells the cell whether conditions are getting better or worse (in terms of food supply) and what should be done about it.

And what does methylation do in molecular terms? Well, it reduces some electrical charges on the receptor and probably makes the receptor less flexible as well. The net result is that the receptor becomes more

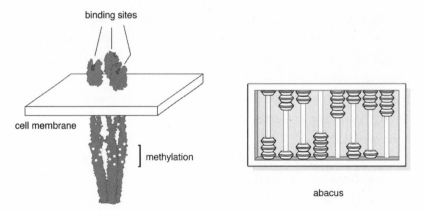

FIGURE 5.2. Receptor methyl groups. A cluster of three chemotaxis receptor molecules is shown on the left. Each receptor has a binding site exposed to the outside of the cell and a long tail immersed in the cytoplasm. The tails carry a number of sites (eight per receptor) to which a methyl group can be added. When the bacterium tastes attracting substances, methyl groups are added—the number reflects how concentrated the attractant is. The methyl groups thus act as a sort of counting device like an abacus (right) that keeps tally of recent exposure to attractants.

likely to adopt its active shape—more likely to stimulate neighboring kinase molecules. Now it adds more phosphate groups to CheY molecules and the increase in CheYp causes the motors to tumble. So protein switching, or allostery, is crucial for this simple form of memory.

A recurring theme of this book is the highly variable nature of cell chemistry. The components of a cell's circuitry exist in many subtly different chemical states—part of the reason they are so responsive, beyond the capabilities of nonliving materials. The cluster of chemotaxis receptors on the surface of the bacterium *E. coli* illustrates this protean quality. If you calculate how many different kinds of chemotaxis receptors there can be—allowing each of eight sites on any of the four receptors to be methylated or not—then the answer is more than one thousand. But if you now consider that receptors in the membrane are actually grouped in sets of three, the number becomes larger. You now have to work out how many kinds of receptor triplets there can be, and

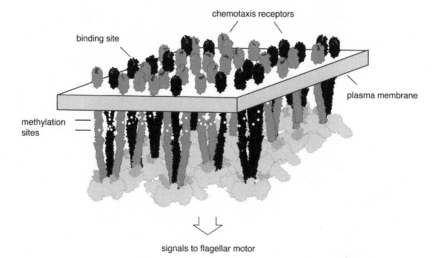

FIGURE 5.3. Schematic view of the cluster of receptors on a bacterial surface. The long thin receptors, here colored gray or black to indicate whether they are active or inactive, are inserted in the membrane with their sensitive binding sites exposed to the outside of the cell. The other ends of the receptors associate with other signaling proteins such as the kinase.

the answer is more than a billion (imagine picking sets of three people from a directory containing more than a thousand names).

But the receptor cluster in any given bacterial cell typically contains several thousand triplets. So the number of possible states it can have as a whole is ludicrously large. It is not only greater than the number of cells in any conceivable population of bacteria; it far exceeds the number of atoms in the known universe. To all intents and purposes, it is infinite.

I'm certainly not claiming that every possible state of a receptor cluster has a distinct biological function. Quite the opposite: most will assuredly be meaningless—just chemical noise arising from the chaotic motions of thermal energy. But because these states exist, evolution could make use of them. Some very small fraction of the different states of the receptor cluster will indeed be important—will cause the bacterium to behave in a different way, even if only slightly and under special conditions. A similar prodigality of other proteins pervades living tissues and underlies their most striking properties, such as immunity and cognition.

Putting things into perspective, I should stress that the events I've been describing take place in a cell about two micrometers long, perhaps a hundred times smaller than the smallest speck you could possibly see with your naked eye. The molecules themselves are hundreds of times smaller again, while some of the important changes such as addition of methyl or phosphate groups take place at the level of individual atoms. Stepping back briefly, it is a triumph that we can know about such an intricate piece of molecular engineering, the more so since my account has been superficial—a broad-brush outline that omits most of the details. A complete description of all the molecular events occurring during *E. coli* chemotaxis would require another, much larger, book.

And it would not be an easy read. As the mountain of data about this simple cell response grows higher, the task of putting it all together in a coherent account becomes ever more daunting. Psychological tests suggest that humans store at most seven short-term memories at the same time and are easily overwhelmed when presented with a long list of interacting processes. It is usually impossible to tell, just by inspection, how even a small set of linked biochemical reactions will work. What will happen if reaction X suddenly increases its rate or if the concentration of Y in the cell falls? Can you be sure that you have all the pieces of the jigsaw puzzle? How do you know that future investigation will not disclose a novel protein or process that entirely changes our view of how the system works?

The logical place to turn for answers to such questions is a computer. Although we poor mortals have difficulty manipulating seven things in our head at the same time, our silicon protégés do not suffer this limitation. Computers have a gargantuan appetite for numbers, digesting them at lightning speed, regurgitating them in any desired form, even presenting answers as pictures that we can understand. It makes sense to give them the task of integrating these lists of interacting biochemical reactions. Supply your computer with the relevant biochemical data, concentrations, rate constants, and so on, then watch the screen!

Curiously, given the many similarities between protein molecules and electronic components, cell biologists have only recently embraced computer simulations. Other areas of biology were much more advanced in this respect. Neurophysiologists were using computers in the 1960s to process their electrical recordings; biochemists in the 1970s were modeling the kinetics of individual enzymes and using computers to estimate the flux of small molecules in metabolism. Structural biology, with its origins in X-ray diffraction and other physical tools in the 1950s, was computer-intensive from the start. Institutes of genomic research, performing today's DNA sequence analyses, are notable for the number and size of the computers they employ. But core problems of cell biology, relating to signaling, movement, control of division, and differentiation, have been more refractory.

Bacterial chemotaxis is a good place to start, since it is generated by a relatively simple set of proteins, all known in some detail. These features have made it an attractive system for computer models. Numerous programs have been written that represent the pathway, with significant success. These show to what extent we really understand how this little pathway works and also reveal discrepancies—cracks in the smooth veneer of established explanation. These discrepancies can lead us to new aspects of the molecular machinery.

The simplest modeling approach, often used in the past to model metabolic reactions, is based on rate equations. Written in computer code, rate equations tell you how fast an enzyme will use up its substrates, and how fast it will make products. Switching of the enzyme glutamine synthase, for example, can be modeled in this way. Rate equations give a first glimpse into how the bacterial chemotaxis system performs. They show how the concentrations of key signaling molecules inside the cell rise and fall when the cell is exposed to different concentrations of attractants and repellents, and what consequences we can expect from different mutations.

In its simplest form, the rate equation method assumes that all of the enzymes and other molecules are freely diffusing inside the cell. More sophisticated methods are needed to display, for example, the receptor complex in the membrane. Protein molecules in the system can be given unique designations (like ID badges) in the computer and their positions in space recorded. Not only their chemical reaction but also their physical movement from one location in the cell to another can be simulated. Another feature introduced by this higher level of graininess is the element of chance. Diffusion is inherently a random process. Whether two molecules bump into each other, and if so whether they bind and react together, has a certain probability. Accurate computer programs have to capture this element of randomness.

Simulation outputs can be displayed in many ways. They can show motors switching between runs and tumbles, for example, or signaling proteins adding or losing phosphate groups. Perhaps the most immediately accessible presentation is a graphic display of simulated bacteria swimming on the computer screen. Bacteria in such an animation can be given sets of flagella that turn and twist like the real objects; the terrain can be colored to reveal user-specified gradients of attractant and repellent. As the silicon bacterium samples its surroundings, changes to its internal biochemistry lead to almost instantaneous shifts in its swimming. Populations of simulated bacteria will then run and tumble across the screen in a realistic manner, swimming toward distant sources of food. Despite many obvious differences from real bacteria, these graphic displays reproduce a broad range of experimental data relating to the sensitivity and rates of response and give authentic swimming patterns for multiple types of mutants.

In a philosophical sense, a computer program representing bacterial chemotaxis or another biological process is a way of knowing—a symbolic representation that helps us to comprehend the phenomenon. This parallel representation of the outside world has much in common with natural language. In his 1960 book *Language and the Discovery of Reality,* Joseph Church, a well-known expert on child development, said that language "permits us to deal with things at a distance, to act on them without physically handling them. . . . We can manipulate symbols in ways impossible with the things they stand for, and so arrive at novel and even creative versions of reality. . . . We can . . . rearrange situations which in themselves would resist rearrangement. . . . We can isolate features which in fact cannot be isolated. . . . We can juxtapose objects and events far separated in time, and space. . . . We can if we will turn the universe symbolically inside out." In much the same way, computer simulations allow us to deal with bacteria without physically handling them. Those lines of computer code capture for us essential features of bacterial movement.

Writing a computer simulation of some process gives you a sense of ownership, like giving a name to something. But both kinds of possession are illusory. For can a computer simulation really include everything there is to know about a swimming bacterium? Hardly! First of all, the sensory range of real bacteria is vastly greater than the attraction to a few amino acids. It is true that a swimming *E. coli* cell has only two forms of motion, run or tumble. But the lengths of its runs and the frequency of its tumbles are sensitive to the concentrations of dozens of distinct chemicals, as well as to heat, salt concentration, acidity, and the metabolic well-being of the cell. And in the real world these are not presented to the bug, as they are in a laboratory, as just one stimulus that increases in a single step or a simple ramp.

Whether it is in the gut of an animal, a sewer, or soil, a bacterium is exposed to an intricate medley of sensations. This symphony contains vital information for bacterial survival, and it would be uncharacteristically obtuse of bacteria not to take advantage . . . surprising if, over the course of evolution, they had not learned that certain combinations signal potential advantage and others impending disaster. The receptor lattice, with receptors of different kinds spatially arranged in multiple variations and subject to subtle modifications, is the obvious location for such discriminations.

Consider the response of *E. coli* to changes in temperature, which it senses through the same set of receptors I have been describing. In sparse cultures where food is abundant, the bacteria are attracted to warmth and will collect in the region of a weak laser spot under a microscope. But in a dense culture where food is limited, the same species of bacteria move away from the illuminated spot: they are now attracted to cold. What possible explanation can there be for such seemingly paradoxical behavior?

The answer appears to be that when food is abundant, it is in the bacterium's interest to grow as rapidly as possible, and growth is accelerated by a move to a higher temperature environment. But if resources are limited, then a smarter strategy is to do the opposite: bacteria move to a lower temperature, "chill out," and wait until conditions become better before growing. In molecular terms, the altered response to temperature is achieved by changing the receptors. Cells in dense cultures inactivate their warm-seeking receptors by methylation and make more of the cold-seeking receptors.

Other examples of environmental response occur when *E. coli* bacteria are cultivated in different conditions. In a nutritionally rich medium the cells stop making their chemotaxic machinery—receptors, signaling proteins, motors, flagella. Why make these when food is abundant? If they are grown on a moist surface, these same cells sprout hundreds of flagella over their surface and swarm in large cohorts, a very different form of locomotion. Individual bacterial cells also send and receive signals to and from the surrounding medium and thereby assess how crowded is their local environment. This is part of the rather poorly understood capacity of some kinds of bacteria, including *E. coli,* to congregate in selective niches. There they adopt a more sedentary mode of existence and secrete large quantities of a sort of mucus that eventually solidifies into a resistant biofilm.

Computer models of bacterial chemotaxis illustrate an occupational hazard for programmers, what you might call reality distortion. People who spend many hours staring at the screen become so entranced at the clever things they see that they begin, ever so subtly, to believe that this is the world itself. I first became aware of this problem when developing models of swimming bacteria. The more realistic my animations became, the more I came, subconsciously, to think of them as the real thing.

There are programs that generate richly complicated images. The equations known as fractals, created by the Belgian mathematician Benoit Mandelbrot, for example, produce wonderfully rich patterns that unfold as you zoom in at higher magnification, revealing unending panoramic views. Another source of highly intricate images are cellular automata, reputedly invented by the Polish-born mathematician Stanislaw Ulam.

Ulam worked at the nuclear research facility at Los Alamos, New Mexico, during the 1940s and was associated with the project that designed and built the first atomic bomb. He also had more playful interests, including the generation of patterns on a computer. Output from Ulam's programs, displayed on punched cards or tapes, evolved in discrete jumps: from moment to moment the fate of any given site, or "cell," depended only on the states of its neighbors. The rules were simple and the technology, by present standards, primitive. And yet these cellular automata had curious, almost lifelike, properties. A simple square would evolve into delicate coral-like growths; two growths might fight over territory, sometimes leading to mutual annihilation.

In his 2002 book *A New Kind of Science,* Stephen Wolfram made a systematic study of some two thousand cellular automata. He codified perhaps several billion more (it is a very large book). Cellular automata are produced by simple rules, easy to state and almost as easily displayed on a computer. And yet some of them are astonishingly rich— strange blends of regularity and randomness. Solid blocks of color suddenly explode into ramifying tessellations like the twigs on a tree. Discrete gems of crystalline order coalesce out of a field covered with a random pepper-and-salt mix. The arbitrary and unexpected nature of these efflorescences inspired Wolfram to make biological analogies, drawing parallels between his cellular automata and the patterns achieved by real cells through evolution and development. He showed automata making shapes like leaves, undergoing sudden transitions reminiscent of embryonic induction, and so on.

Reality distortion indeed! These computer-generated spatial patterns or temporal sequences are nothing more than superficial ciphers. A single leaf of grass is immeasurably more complicated than Wolfram's entire opus. Consider the tens of millions of cells it is built from. Every cell contains billions of protein molecules. Each protein molecule in turn is a highly complex three-dimensional array of tens of thousands of atoms. And that is not the end of the story. For the number, location, and particular chemical state of each protein molecule is sensitive to its environment and recent history. By contrast, an image on a computer screen is simply a two-dimensional array of pixels generated by an iterative algorithm. Even if you allow that pixels can show multiple colors and that underlying software can embody hidden layers of processing, it is still an empty display. How could it be otherwise? To achieve complete realism, a computer representation of a leaf would require nothing less than a life-size model with details down to the atomic level and with a fully functioning set of environmental influences.

Returning to the simple protein circuit that is the subject of this chapter, we have one further lesson to learn. When I described the cascade of reactions in a bacterium responding to an attractant, I made a huge simplification. Each step of the pathway stood for a particular *type* of molecule. So if the cell had 8,230 molecules, say, of the kinase associated with the receptors, I bundled these into a single mathematical construct—one node of the network. One benefit of this approach was that of exactness. I could then calculate how that ensemble of identical enzymes would behave in the next second—precisely how many phosphate groups it would add. Biochemists use this kind of approximation all the time. The field of enzyme kinetics is predicated on experiments performed on large volumes of aqueous solution (large, that is, compared with a single cell) containing an effectively infinite number of uniformly dispersed molecules. This allows biochemists to write mathematically precise equations for the chemical conversions.

But conditions in a real cell are not like this. Living cytoplasm is a curious substance unlike anything usually considered by physical chemists. Molecules are present in a slurry rather than in solution. There is a great deal of organization and spatial variegation. Location is crucial, just as it is in a crowded metropolis; next-door neighbors can change your life, in the world of proteins as in a condominium. You might be rubbing shoulders with an enzyme that modifies you chemically, or a protein that transmits a conformational change. You might be sitting next to a mitochondrion that floods your environment with ions and small molecules. A motor protein might suddenly appear on the scene and deliver a package of immigrant enzymes. In other words, our population of 8,230 kinase molecules is far from uniform. It is made up of distinct cohorts in different locations, each able to behave slightly differently.

Evidently it is a gross simplification to represent all kinase molecules in a cell as a single mathematical cipher. Ideally one would treat each molecule individually, include appropriate features of its particular location and surroundings. But a transition of this kind requires more than just a multiplication of factors: it also calls for a fundamentally different approach. For any individual molecule is intrinsically capricious and unreliable, dominated by thermal energy. Molecules not only perform a dance of undirected motion but also undergo unceasing turmoil inside. You can never know with complete certainty whether a particular enzyme molecule will be attached to its substrate or precisely when its many atoms will perform catalysis. Strictly speaking, all you can say is that there is a certain probability that a certain reaction will occur. If you average this probability over a large enough number of molecules (more than ten thousand is usually enough), then the variations from molecule to molecule are ironed out.

The random, noisy behavior of biological molecules is not just a technical nicety. It determines the fate of entire cells and organisms. In 1975 John Spudich and Dan Koshland watched swimming bacteria under a microscope and saw that each cell had its own characteristics. Lewis Thomas later described their results in one of his charming essays: "Motile microorganisms of the same species are like solitary eccentrics in their swimming behavior. When they are searching for food, some tumble in one direction for precisely so many seconds before quitting, while others tumble differently and for different, but characteristic, periods of time. If you watch them closely, tethered by their flagellae to the surface of an antibody-coated slide, you can tell them from each other by the way they twirl, as accurately as though they had different names."

This is biological individuality. These bacteria had all grown from the same colony. They were to all intents and purposes genetically identical. And yet, through the intrinsic noisiness of their molecular processes, each cell displayed a distinct pattern of swimming. One reason for this is that the receptors and signaling molecules in the signaling pathway were present in slightly different numbers from cell to cell.

I've already mentioned that the information for the structure of proteins is carried in DNA. But because DNA is made of a chain of nucleotide bases, whereas a protein is made of amino acids, there has to be a translation. An elaborate apparatus exists in all cells that makes protein molecules according to the instructions in DNA. The details are complicated, but in broad terms a limited region of DNA—one gene—is copied into a RNA molecule. RNA is similar to DNA, with some minor chemical differences (it is made of ribonucleic acid as opposed to deoxyribonucleic acid). The RNA then moves from DNA to the cytoplasm, where it directs the synthesis of proteins. This it does by threading through large particles called ribosomes that move along the RNA spooling out a new protein molecule. The flow of information in a cell therefore passes from DNA to RNA to protein, at each stage being encoded by a linear sequence of subunits in a polymer. The rubric *DNA leads to RNA leads to protein* is sometimes referred to as the basic dogma of molecular biology.

Working flat out, a ribosome might add ten amino acids per second to a growing protein. A typical protein might require a minute or so to be made. But as with every other aspect of cell chemistry, there are numerous checks and balances. The complicated stream of biochemical events, from DNA to RNA to ribosomes and eventually to protein, can be controlled virtually anywhere along its length. Regulation entails interactions with other protein molecules and is subject to the inexorable laws of thermal diffusion. So the number of copies of this protein in a cell is subject to the laws of chance: the toss of a coin.

That is why the bacteria observed by Spudich and Koshland showed such large differences in their individual swimming patterns. Each cell contained the same proteins but in slightly different numbers. The resulting subtle differences in composition were enough to make runs of an individual bacterium longer or shorter, its tumbles more or less frequent.

And what is true for a small bacterium is even more apparent in plants and animals. Individual variation is a hallmark of biology; no two organisms are exactly the same. No two cells are exactly the same. Part of the variation is genetic, the processes of mutation and assortment of genes creating diversity in the genetic material. One person's DNA is as different from the next as is the bar code of one product from that of the next item on the supermarket shelf. But even if every effort were made to eliminate genetic differences, cells and organisms would still not be identical. A cloned cat has a different coat pattern from its genetic mother and a distinct personality. Human identical twins have different fingerprints and different iris scans.

You might say at this point, so what? Even inanimate objects differ to some degree; no two pebbles on a beach are exactly the same. Surely if I looked closely enough I would find small differences even between two microchips made from the same silicon crystal in the same clean room facility? True: but what sets biological systems apart is the multiplicity of ways they can be different and how many of these differences have really important effects. Even a single event at the atomic level, amplified by cascades of reactions, can result in a radical change at the macroscopic level. A young man develops sarcoma of the leg and dies within a year. A tiny accident during embryonic development creates a musical genius, another a child with a harelip.

Consider the influence of noise in your genetic makeup. Every one of the zillion cells in your body carries twenty-three pairs of chromosomes. One chromosome of each pair came from your mother (more precisely, it is a copy of one supplied by your mother) and one from your father. When your mother made the egg with your name on it, she donated to it twenty-three of *her* chromosomes. When your father made the sperm cell destined to fertilize your egg, he did the same. And how did they choose which chromosome of each pair to give you? Well, they didn't: it was decided randomly, by molecular noise. During the developmental process leading to egg and sperm, members of each pair of chromosomes are segregated into different daughter cells. This is a mechanical process, managed by microtubules. And it occurs in a truly random manner, with pairs of chromosomes spinning like coins before they land in one or the other position. Consequently, each sperm and each egg contains a highly distinctive set of chromosomes. The combination of the two produced by fertilization, even from the same parents, is an essentially unique combination of about 1 in 70 trillion chromosomes.

SIX

Neural Nets

"Gug gah, bah bit," the childish voice resonated in the elegant space of the Lady Mitchell Hall, a large lecture theater built for the University of Cambridge in the late 1950s. "Doo doo, wha woo" . . . an infant's babble projected to an audience of perhaps three hundred academics, listening in solemn attention. The recording came from a laptop computer on a table, beside which sat a slim man in a dark suit, silently confronting the sea of faces before him. As we listened, the string of nonsense continued. What were we listening to? Then, unexpectedly . . . Did I detect a real word? Not "wha," surely, but "where"? Was that "boo" or "blue"? Abruptly, and with that sharp transition that sometimes comes when listening to someone with a foreign accent, I realized that the voice from the laptop was speaking English. Distorted in a strangely chaotic manner, but English nonetheless. Slowly it became clearer and easier to understand. Within a minute or so, we knew that we were listening to the recitation of a child:

> When I come home from school I uh change my clothes and uh
> I play with my friends and five o'clock we have to go pick up
> Daddy and then we come home and eat supper but first we feed
> our rabbit for his supper.

It was not, in fact, the voice of a real first-grader holding the attention of the assembled academics but of a computer program called NETtalk. The occasion was the annual Darwin Lecture series of 1996, sponsored by Darwin College, Cambridge. We were midway through one of the lectures, listening to a record of NETtalk learning to pronounce English—the process having been speeded so that it took minutes rather than hours to play. The program had been codeveloped by that evening's speaker, Terry Sejnowski, professor at the Salk Institute in San Diego. In fact, it was a special kind of computer program called a neural network, based on a web of interacting elements loosely modeled on brain cells.

The network had been set up initially in a purely arbitrary manner as a nonspecific, generic set of connections. It had an input and an output and could be trained to match one to the other. For this demonstration, the network had been given a string of letters and spaces corresponding to a passage of English text from a five-year-old and had to produce a string of sounds from the written text, as an English speaker would have done. Not a trivial matter, since English is an especially difficult language to pronounce because of its irregular spelling.

The way the network achieved this feat was indeed counterintuitive—almost spooky. For there was in fact no voice coach, no elocutionist. No one showed the software how to match this or that sequence of letters to this or that sound. All the programmer did was provide the equivalent of crib sheet, a test list of words with their correct pronunciation, like a Rosetta stone. Before each test the network made small, tentative changes to the strengths of its connections; afterward it received a score according to whether its performance had improved or deteriorated. Networks with the best scores were retained for refinement, whereas those with poor scores were discarded. Like a child learning by trial and error, the program then found its own way to a solution. It learned the regularities of consonants and vowels before the irregularities. It went through distinct phases of babbling—got simple words first before reaching a respectable and understandable cadence. To judge by the silence of the audience, most listeners found the process uncannily realistic. It was strangely similar to a human baby progressing from babble to speech.

Neural networks belong to the field of artificial intelligence. They arose from the notion of artificial nerve cells, introduced in Chapter 4, and specific implementations such as that of perceptrons—introduced by Frank Rosenblatt, a neurobiologist at Cornell University in 1958. A perceptron is a computational device, roughly modeled on a real nerve cell or a small portion of the nervous system. It receives a set of inputs from outside, calculates a weighted sum of these inputs, and then produces one of two possible output values: yes or no.

Initially there was great interest in using perceptrons and similar devices to reproduce and explain the operation of parts of the nervous system. But it quickly became apparent that real nerve cells are more sophisticated in their responses and wired in much subtler ways. What proved to be more lasting and productive were the collective properties of systems of perceptrons. Sets of perceptrons linked in defined ways on a computer provided the source of much fascination to theoreticians—and unexpectedly also had numerous practical applications. Today, the descendants of these systems, neural networks, are universally used in pattern recognition, from the recognition of faces and handwriting to the playing of chess and backgammon, from diagnoses of medical conditions to the filtering of email spam.

I first heard about neural networks from Jerry Pine, when he visited my lab at King's College, London, in 1987. Jerry—professor of physics at Caltech and an enthusiast on every subject—is a walking encyclopedia of the physical world. Certainly, at the time, he seemed to us to know everything—about how fluids behave in very small spaces, how light and electrons interact with matter, how computers work. He gave informal talks in his soft American accent on the late-breaking news from the West Coast, including exciting stories about neural networks. What most caught my imagination about these devices, as Jerry described them, was their autonomous nature. They could be trained to perform sophisticated tasks of recognition or discrimination without knowing the underlying rules. Neural networks often achieve a level of sensitivity and discrimination no human can match. To an outside observer, the learning appears as if by magic. Of course, if you wish you can play the detective: take the network apart and deduce, after the fact, how it achieved its feat. But the process is autonomous: no human intervenes to say that you must make this or that change to the network to achieve the learning.

Researchers have invented different routines to train neural network programs, some based on simple trial-and-error, others using sophisticated mathematical procedures. But the basic strategy is always the same. Essentially the network is presented, again and again, with a small test set of inputs. These may be of almost any kind: they may be amino acids from a protein sequence, pixels from a picture of a face, or signals from an electronic nose. In the case of NETtalk, the inputs are sets of letters from a passage of text recorded live from children's speech. Once the test set has been applied, the computer then calculates how close the answers are to the correct ones. NETtalk, for example, compares the output with recorded speech converted to a stream of sounds, or phonemes.

Then, having obtained a score for how good this particular network is, the computer tries a variation. It makes a new network based on the previous one by changing the strengths of some of the connections. In the simplest method the changes are made at random, making one connection stronger, another weaker and so on. This method has analogies to natural evolution and has direct biological implications, as I will show. Other more complicated strategies exist, but the outcome is always selected on an empirical basis. At every cycle, the computer makes changes and tests the new network. If the result is an improved performance (closer to the target) the new network is retained. If it is worse, it is discarded, and the computer tries again.

A computer-based neuronal network is like a caricature of the nervous system. Each unit resembles a nerve cell, in that it can adopt either an active or an inactive state, comparable to the electrical firing of a nerve cell. An active unit transmits its state to neighboring units through its connections, a bit like the long threadlike extensions called axons that link one nerve cell to another. Whether or not a particular unit fires depends on the total of all of its inputs, added according to some predefined rule. This might be as simple as, for example: IF THE SUM OF INPUTS IS GREATER THAN ONE THEN BECOME ACTIVE. Learning in the brain depends on the modifiability of connections between cells called synapses, which become stronger or weaker with experience. In place of synapses, the computer-based connections have a variable weight that controls the size of the signal that passes to the next unit.

NETtalk belongs to the class of feed-forward networks that operate in a single direction. Training consists of a series of repeated pulses or waves of activity spreading from input to output. As a wave reaches each individual unit (analogous to an individual nerve cell or neuron), its activity is recalculated on the basis of the connections it receives from preceding units. These in turn depend on the activities of preceding units each multiplied by the weight of the connection. Weights change with each updating sweep, and the entire process is repeated until the network achieves the desired level of accuracy (or fails to find an answer, as sometimes happens).

Once a network has become trained, its weights and activities acquire particular values that make the network function correctly. You might say that the network has then "captured a representation of input patterns," or, more provocatively, "acquired knowledge of its environment." Thus, in the case of NETtalk, a layer of input units encodes the letters fed into the network. Each letter is assigned to one of a cluster of twenty-nine input units—one for each letter of the alphabet plus an additional three units to encode punctuation. At the output end of the network a set of twenty-six output units represents the phoneme assigned to each letter. As a group, they encode such features as articulation, voicing, stress, and syllable boundaries and can be sent to a loudspeaker for broadcast.

But the meat of the sandwich is in the middle: the intermediate layer between input and output. Units in this layer—NETtalk had eighty of them—are termed hidden units because the programmer does not control them. At the beginning of the training period, starting with a naïve network, hidden units are assigned arbitrary, random activities. As the network learns the assigned task, each connection of a hidden unit with its neighbors slowly acquires a certain weight. Then, when the trained network is offered items from the test set, activities of hidden units rise and fall in a reproducible fashion with the input. But what are these values? What do they represent?

When Sejnowski and Rosenberg examined this question in the case of NETtalk, they found that many different sets of weights gave about equally good performance; different networks solved the same problem using different sets of connection weights. Moreover, the efficiency of a network and how accurately it learned changed with the number of hidden units. The more hidden units, the richer the performance—at least within limits. Sometimes it was possible to understand the function of a particular hidden unit—for example, it might have been strongly associated with a set of letters or with certain phonemes. But the most striking regularities involved groups of hidden units that became active at the

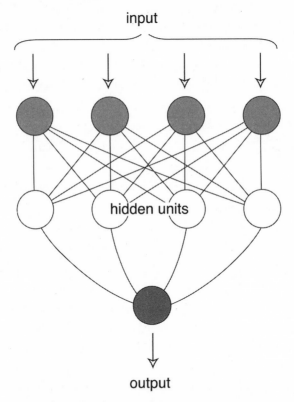

FIGURE 6.1. A simple neural network. The computational units (circles) are arranged in three layers, with each unit in one layer connected to every unit in the neighboring layer. The programmer trains the network by applying a set of stimuli to the input layer and adjusting the strengths of connections until the output unit gives the desired outcome.

same time. Some of these groups could be readily identified, such as those distinguishing vowels and consonants.

Something like this occurs in the human visual system. Nerve cells in the retina produce signals that depend on the pattern of light entering the eye—they perform a function comparable to the input units of a neural network. At the other end of the system, projections to other parts of the brain trigger appropriate responses to visual stimuli—these are like output units. And the hidden units in this analogy are the many neurons at intermediate locations of the brain that process the raw input. The visual system is a good choice because a great deal is known about the processing at these intermediate stages. Close to the visual stimulus, individual neurons recognize primitive features of the visual scene, such as orientations and edges. But farther downstream from the eye and closer into the brain, nerve cells become more specialized and sophisticated. They start to encode such modalities as motion, color, and binocularity.

Almost fifty years ago, the concept of increasing specificity was epitomized by the term *grandmother cell,* referring to a notional nerve cell in the visual system that responds optimally to an image of one's grandmother. There is now good evidence that cells in the visual system indeed show highly selective responses to particular visual stimuli. The task presented to a neural network such as NETtalk, although very much simpler and less demanding than the tasks the visual system has to tackle, is solved by an analogous procedure. In both cases there is a progressive integration of perceptually significant features as input segues into output.

How does this integration happen? Every unit in a neural network receives inputs from what went before, so at the very least what it represents will be a summation of its inputs. But it might be more complicated than this. For example, a unit buried in a neural network might be subject to a rule such as: BECOME ACTIVE IF INPUTS FROM A AND B ARE STRONG BUT NOT IF INPUT IS RECEIVED FROM C, or its instructions might be of this kind: SEND A SIGNAL ONLY IF THE SUM OF INPUTS FROM A AND B

IS GREATER THAN THE SUM OF INPUTS FROM C AND D. The appropriate term here is *combinatorial:* each unit represents some combination of the upstream units feeding into it.

Now imagine moving to the next layer. Units in this layer receive inputs from the previous one, exactly as before. They will again combine these inputs according to some mathematical rule. But the inputs with which they start are different—already partly digested. So the same rule of integration necessarily produces a more complicated output because it is a combination of combinations. And as you move through the network, this same process will continue. At every step, units become less like the original input to the network and more like the final output.

I'm not talking here about fragmentation into a thousand pieces like a laser hologram. Rather, the features recognized by a trained network become distributed among units operating in parallel. Typically, a group of units in a trained network will give their strongest response to a general feature (such as vowels or consonants). If you remove just one connection or hidden unit from this group—a surgical procedure that is painless on a computer—the effects are rarely dramatic. Usually one observes a decrease in the level of performance: a poorer distinction between different inputs, a blurring of the edges. When hidden units were dissected from the NETtalk, for example, the program did not suddenly forget how to pronounce th or a soft c. Instead, its performance became generally of lower quality, the output speech increasingly indistinct and slurred.

The charming phrase used to describe this fault-tolerant process is *graceful degradation*—an aspiration for everyone over the age of fifty! And the complement of graceful degradation (elegant recovery?) also applies, at least for computer networks. If the damaged portions of a network are repaired, then recovery of function quickly follows, building as it does on a scaffold of undamaged connections.

Neural networks were invented to be, literally, models of the nervous system. But with a few exceptions (such as the olfactory system), their usefulness in this regard has been limited. Real nerve cells are usually too idiosyncratic to be thought of as standard units that act in a stereotypical way. Protein molecules, though, are a better fit. I've already argued that an enzyme in the cytoplasm or a receptor in the membrane is a logical element similar in some respects to a transistor. The protein might read in the concentration of one or two small molecules and change the concentration of another. Moreover, in most if not every case, a regulatory molecule controls the enzyme or receptor. Each protein therefore has multiple inputs, performs some logical operation, and generates an output. It is equivalent to a connection point, or node, in a neural network.

I also have argued that simple circuits are built from protein molecules, using as an example the chain of signaling reactions in bacterial chemotaxis. This pathway is small and relatively independent of other processes . . . but the operative word here is *relatively*. You just have to scratch the surface to uncover a multitude of links to all kinds of other cell processes. The kinase that drives the cascade of chemotactic signals, for example, also plays a part in other signaling pathways, such as responses to osmotic changes and the detection of glucose. Just making the proteins of this pathway and positioning them in the membrane and cytoplasm requires numerous other molecular interactions to be performed. Following this line of thought, I must admit that ultimately I should be able to trace connections between chemotaxis and every other function of the cell: no circuit is an island unto itself.

To see how interconnected everything is in a cell, consider the response to a single hormone. During a period of fasting, blood levels of a hormone called glucagon rise—a signal to the body that its supply of fuel is running low. Molecules of glucagon travel through the body from where they are made (the pancreas) to where they are used (the liver). Binding tightly to specific receptors on the surface of the liver cells, glucagon molecules cause the receptors to change shape. On the cytoplasm side of the membrane, the receptors trigger a cascade of downstream signals and eventually cause the release of glucose from granules of the storage form glycogen. Glucose molecules spill out of the liver cell into the bloodstream to be carried throughout the body.

The signals just mentioned that travel from the glucagon receptors in the membrane to the glycogen granules in the body of the cell are carried by many different molecules. But perhaps the most crucial player is the small molecule cyclic AMP, which rises rapidly in concentration shortly after the liver cell has tasted glucagon. This happens because the chain of reactions from the receptors has activated an enzyme, adenylate cyclase, that manufactures cyclic AMP from the ubiquitous material ATP. Farther downstream, the sharp rise in cyclic AMP triggers other changes, many of them entailing the addition or removal of a phosphate group. At the end of this chain an enzyme adds a phosphate group to a crucial enzyme, glycogen phosphorylase. This modification switches the enzyme to a form that degrades glycogen to produce glucose. Thus, through a chain of perhaps a dozen distinct protein modifications, the hormone glucagon triggers release of glucose in the liver cell.

Every protein in this chain also responds to molecules outside the pathway. Cyclic AMP levels depend not only on adenylate cyclase but also on enzymes (phosphodiesterase, or PDE) that degrade cyclic AMP, each regulated in a distinct fashion. The last enzyme in the pathway, glycogen phosphorylase, is another example of complicated enzyme regulation, being controlled by multiple kinases in turn controlled by different cell processes. Figure 6.2 also shows why binding of glucagon to the surface of a liver cell has multiple effects. It stimulates not only a rise in cyclic AMP (and hence glucose) but also important changes in specific lipids in the membrane. These in turn trigger multiple changes, including release of calcium ions from internal stores.

So a liver cell's response to a single hormone, glucagon, is far from being a linear chain of cause and effect. It is a pattern of signals that spread outward from the initial binding of glucagon to their receptor on the cell surface. Multiple changes travel down parallel pathways before converging onto a specific molecule such as glucose. Moreover, other hormones such as insulin and adrenaline—and even metal ions, electrical signals, and mechanical stimuli—influence the same signaling molecules. A change in glucose will thus be produced by a plethora

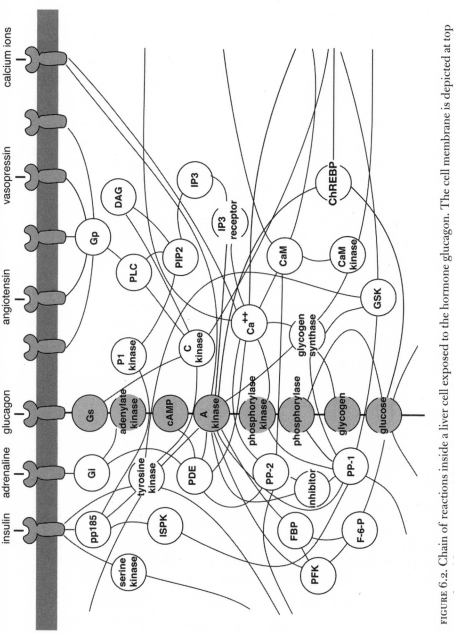

FIGURE 6.2. Chain of reactions inside a liver cell exposed to the hormone glucagon. The cell membrane is depicted at top together with receptors and the substances such as glucagon they bind to. Below the membrane, signaling proteins are shown as computational elements similar to those in a neural network.

of external conditions; the concentration of glucagon is just one of many factors.

The biochemical pathways regulating glucose in a liver cell make up a sort of neural network. The inputs on that network include glucagon, insulin, adrenaline, and so on, each binding to and activating specific receptors in the cell membrane. Depending on the connections of the pathways and their weights, these multiple inputs will be combined. They will be processed as the signals that they trigger spread throughout the system. What you consider to be the output is arbitrary, since there is no well-defined endpoint. But if, say, you choose the concentration of glucose inside the cell as the readout, then this value will fluctuate in systematic ways with different inputs.

Of course, this is a neural network with special features. First, the units are not identical in design, as they are for the computer-based variety: each unit corresponds to a different kind of molecule and therefore requires distinct inputs and outputs. Second, the pattern of connections of a cell network is much less regular. Units are not arranged in well-defined layers, and some signals pass backward as well as forward. As a consequence, the output of the cell network cannot be calculated in one single pass and has the potential of generating cyclic or other temporal variations with each round of calculation.

There is also the question of training. Before a set of interconnected biochemical reactions can function as a neural network, it has to be educated, its connections modified by experience. A computer neural network changes its connection strengths in response to a teaching input. A training set of inputs is presented to the network again and again, and the output compared with the correct answer. The difference is then used to modify the weights of the network. In one common procedure, the error is used to modify weights working backward from the output layer to the input layer. After each backward pass, the network is again presented with the training task and the error is again measured. The process is continued until the error has dropped to an acceptably small value.

What would be the corresponding training regimen for a cell? To learn a new task, a network of proteins has to change its connections. That is, it must alter binding strengths—change the rates of catalysis, modify the efficiencies of regulation—all in response to external influences. How could this happen? This is indeed a crucial question and one that goes to the heart of the notion of wetware. The answer depends, I will argue, on the ability of the molecules of living organisms to change according to their surroundings. They change their chemical structure—I will use the term *morph*—so as to facilitate the survival of the organism. Familiar examples of this process occur in the brain, as you store new memories and learn new tasks, and in the immune system, when your body develops antibodies to resist the onslaught of a new infectious organism.

But the most far-reaching and important mechanism that cells use to train networks of proteins is evolution. The performance of a network depends on the processing capacities of its protein elements—the way they combine inputs, how much weight they give to particular connections. These in turn depend on protein structure, specified by sequences of nucleotide bases in DNA. The seed corn of evolution is the continual variation in base sequences produced by mutation and rearrangements of the genetic material. From this long-term viewpoint, each learning trial corresponds to a single generation; each "test" is the ability of the organism containing the new network to survive and reproduce.

The notion that protein networks might evolve seems strange at first. Most people are familiar with the idea that individual genes can change with time. But I am now asking sets of genes to change at the same time, to work as a group in order to produce an improved collective performance. But in fact, similar strategies are widely employed for computer-based neural networks, and they can be highly effective at training.

In 1994 my friend Steve Lay and I wrote a program to examine how a simple signaling pathway might evolve. Actually, it was more like the selective breeding of an animal, because it avoided all the complicated gene-shuffling population effects of true evolution. The pathway we set up was an idealized signal cascade of a cell responding to a substance in its environment. A hormone bound to a receptor in the membrane and triggered a change in a protein inside the cell. The signals were carried, as in bacterial chemotaxis, by a cascade of molecular changes. Receptors changed shape when they bound the hormone and became active. The active receptor changed the rate of addition of a phosphate group to a target protein in the cytoplasm. The net result was that addition of hormone on the outside of the cell (the input) caused an increased number of phosphates to be added to the signal protein (the output).

The one twist we made to this storyboard—what made it interesting— was that we gave our model *two* types of receptors. Both bound to the same hormone and both fed the same signal cascade. But their binding and enzymatic properties could change independently. The input-output performance of the network then depended on seven independent parameters, relating to how strongly different molecules bound to each other and how efficiently the enzymes worked. With the values we started with, an on-off pulse of hormone produced a matching transient change in target protein: it was a simple relay.

Then we introduced the training regimen. If we made small random changes to one of the seven parameters of the network, its output would change. Given a standard input pulse of hormone, the output pulse of signal phosphate would change its size and shape. Because this was a computer model, it was a simple matter to do this repeatedly, sprinkling random changes onto the network. We thereby created a family of offspring networks, each a descendant of the first. Each of these slightly altered offspring was then allowed to produce its own descendants; at each step the progeny diverged further from the starting network. Proceeding in this manner, we built a family tree of networks. The process resembled reproduction with variation, one of the essential ingredients of evolution.

We then applied the other essential ingredient: selection. After each family of networks had been produced by mutation, we tested its function. In one experiment, for example, we required the networks to invert the input. They had to turn a positive pulse of hormone into a negative pulse of target protein. In another experiment we asked the networks to convert a transient pulse into a permanent change in intra-cellular signal, in effect working like a toggle switch. We scored the net-work in each family after successive rounds of mutation to find the one that came closest to our desired objective. The winner was then used as the founder of a new dynasty, and the latter, in turn, subjected to other rounds of mutation and selection. Progressing in this manner, we found that our networks became increasingly better at the required task. In bi-ological terms, they bred closer and closer to the desired traits.

It was surprisingly easy to shape the networks. Output traces started to change after just a few rounds of mutation. Not always in an obvious manner—the pulse of target protein would often grow or shrink in size

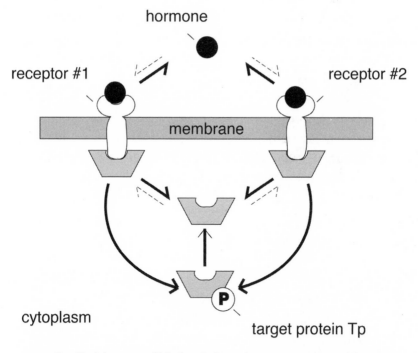

FIGURE 6.3. Training a set of biochemical reactions on a computer. A hormone binding to two receptors on the outside of a cell sends a signal into the cytoplasm, measured by the amount of a target protein. The intensity of this signal depends on the amount of hormone and the seven reactions indicated by black arrows. The computer program adjusts the strengths of these reactions until the output of the network changes in the desired manner.

first and only later alter its shape. But for simple tasks such as inverting the pulse from positive to negative or making it last for a longer duration, it got there in the end. Perhaps the most intriguing result came when we trained the network to respond optimally to a certain strength of hormone. That is, we asked it to give a large output when presented with a pulse at one concentration but little or no output with pulses much higher or lower than this. Our little network learned how to do this in a way that was both unexpected and instructive.

As the training sessions proceeded, the binding strengths of the hormone to the two receptors started to drift apart. One receptor became high-affinity, such that even a small trace of hormone would stick. The other receptor became low-affinity, requiring high amounts of hormone before it could respond. Moreover, the two receptors came to work in different ways: the high-affinity receptor produced a positive output, whereas the low-affinity receptor produced a negative output. Working together, the two exerted a push-pull control over the phosphate cycle that gave the highest response at the target concentration.

Steve and I later found that some biological systems behave in fashion similar to our program, despite its rudimentary and artificial nature. The receptors for many growth factors and hormones often exist in both high- and low-affinity forms. In some instances the two forms enable the cell to respond optimally to a specified concentration, as in our

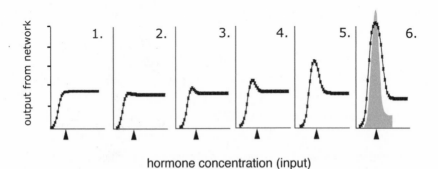

FIGURE 6.4. Results of training. In this series we selected networks that gave the best response (produced the highest concentration of the target protein Tp) when they were provided with hormone at a specific concentration (indicated by the arrowhead). Successive training runs showed a steadily improving performance.

simulation. It seems possible that they could have evolved by a closely similar process to that of our silicon-based networks.

The program Steve Lay and I wrote is a homespun version of a computational approach known as a genetic algorithm. Here the aim is to exploit biological principles to solve difficult problems, such as designing aeronautical wings or predicting the performance of stocks. The input to a genetic algorithm is a set of possible solutions to that problem represented in strings of code, sometimes fancifully termed a "chromosome." These candidate solutions, often generated by a random process, are then tested to see how well they perform a given task; in our case this was how concentrations of the target protein changed. Candidates that fail to work well are deleted, leaving more successful chromosomes to reproduce. Multiple copies are made, but with random changes introduced during the copying process (loosely modeled on mutations and genetic crossovers). These digital offspring then go on to the next generation, forming a new pool of candidate solutions, and are subjected to a second round of fitness evaluation, and so on. Since the average fitness of the population increases with each round, repeating this process for hundreds or thousands of rounds can lead to the discovery of very good solutions to the problem.

Returning to the cell, if sets of biochemical reactions indeed have some of the properties of neural networks, what consequence will this have? Perhaps the most obvious implication is that cells will respond optimally to *combinations* of stimuli. Research biologists like to perform experiments under squeaky-clean, tightly defined conditions. They dissect cells from the organism and grow them in defined nutrients. Keeping everything else the same, they gingerly vary just one component—a single nutrient or growth factor—and watch what happens. It is a reductionist strategy that has been hugely successful—the wellspring of molecular biology. But real life is not like this. Most cells experience a riot of different salts, sugars, amino acids, lipids, hormones, growth factors, and signaling molecules from other cells. These substances fluctuate in concentrations over scales ranging from minutes to days. On the few occasions that investigators have tried, contrary to usual practice, to vary more than one component at a time, they have uncovered unexpected interactions. Published experiments show that certain combinations of growth factors have far greater effects than any of the ingredients individually.

From a logical standpoint, this is just what you might expect. It arises, simply, from the architecture of the network—because it is highly interconnected. Say you have five different receptors in the membrane that all affect a single target protein in the cytoplasm. Say they control the rate at which the protein acquires a phosphate group. Then it is obvious that the concentration of phospho-protein will rise and fall in response to what the five receptors see. All that target protein knows about the outside world is what it is told by the five receptors.

And the messages they send might be highly biased and one-sided. In the simplest case, the phosphate-carrying target protein will simply rise and fall with the sum of the five signals. But if the connections are of different strengths, as is usually the case, then the signaling protein will be more difficult to predict: it will carry a more intricate representation, perhaps flagging just a few combinations of the possible inputs.

In general, you can say that signaling proteins, working individually or as a small group, resemble the hidden units of a neural network. That is, they have a semantic content, or meaning. In the case just mentioned they would convey a set of environmental stimuli to the cell. What this set means in biological terms can be hard to express in words, but not always. There are proteins that tell a cell to adopt an elongated shape or migrate. A secreted protein triggers an embryo to make hair cells; another protein stops a cell in its tracks and makes it commit suicide.

There is a protein found widely in plants and animals that acts rather like a low-battery warning. The enzyme AMP-kinase plays a crucial part in the energy balance of cells, stimulating energy-generating pathways and inhibiting processes that consume energy. This is yet another switch. Its principal regulation is by ATP, the primary source of chemical energy of the cell. Low levels of ATP turn the enzyme on. But in true hidden-unit style, AMP-kinase receives many other signals via other connections that add further levels of control. It responds to levels of the glucose-controlling hormone insulin and the appetite-regulating hormone leptin;

it is activated by exercise and by flight-or-fight signals from the nervous system. So by integrating multiple nutritional and hormonal signals, this enzyme acts as a sensor of cellular energy status.

Consider another enzyme, abundant in muscles of the gut. Myosin light chain kinase adds a phosphate group to myosin and thereby drives the physical process of contraction. Addition of the phosphate not only switches myosin to an active state but also changes its location in the cell. In this way the cell builds a highly organized and efficient protein machine to sustain contractions. From all that has been said, you might expect this crucial switch to be highly regulated itself. And yes, myosin light chain kinase sits like a spider at the center of an entire web of molecular connections. These include signals from the nervous system as well as hormones and other regulatory molecules arriving at the muscle cell membrane via blood circulation. These inputs rise and fall depending on whether you are relaxed or tense, hungry or replete, tired or full of energy. Filtering down through the web of connections, a medley of signals evokes an appropriate response in your myosin light chain kinase. This single enzyme signals to muscles in your intestine, blood vessels, and sphincters when to contract, and by how much.

I've argued in this chapter that computer-based neural networks illuminate features of reactions in living cells. They show us why individual proteins can respond to combinations of events: it is an inevitable consequence of their connectivity. And related to this principle is their ability to detect salient features of their input. After a neural network has been trained, certain (hidden) units respond optimally to particular kinds of patterns of input, such as vowels and consonants in English text. The corresponding notion for a living cell, I suggest, is that of a "semantic protein," with examples including proteins that flag the onset of cell death, act as a fuel gauge for the cell, or control contraction of the gut.

One other feature of neural networks is relevant for us. I mentioned that a network trained to recognize a set of patterns is surprisingly resistant to damage. For example, if a programmer makes random changes to the strengths of connections, these changes usually do not erase whole sectors of the performance. The most common outcome is that the overall level of performance becomes slowly degraded. This property of robustness, arising from the distributed nature of the connections, is also relevant to living systems.

Bé Wieringa is a cell biologist at the Center for Molecular Life Sciences in Nijmegen, The Netherlands. He heads a group of researchers interested in the flux of energy-rich molecules in mammalian cells, especially their extensive interconnections—what you and I would call their "network" properties. One of Wieringa's scientific loves is creatine kinase—an enzyme that catalyzes production of creatine phosphate, a small molecule used by many tissues as a convenient backup source of ATP. One type of creatine kinase is especially abundant in the brain, where it supports the high-energy demands of this tissue. So when Wieringa and his colleagues decided to genetically engineer mice lacking this enzyme (so-called knockout mice), they expected dire results. Without an adequate supply of energy, nerve cells quickly die—so what hope would these mutant mice have?

The astonishing answer was that the mice were fine, thank you! Knockout mice were born, developed, grew to a good size. They were fertile, and, so far as one could tell, every bit as healthy and active and long-lived as their normal relatives. Surprised and, one suspects, a little frustrated, Wieringa and his colleagues kept looking until at last, using sophisticated tests and harsh conditions, they found some differences. In high-resolution images of the brains of the mutant mice obtained by electron microscopy, they saw subtle changes in the number of nerve cell processes in a particular part of the brain. The mutant mice did somewhat less well in finding targets in a water maze; their response to sudden sounds was less acute.

How could this be? How is it possible to remove a keystone of brain energy production and yet leave the edifice intact? The puzzle grew as Wieringa and his colleagues examined other energy-generating enzymes in the same manner. In every case the knockout mice looked superficially normal and they needed sophisticated tests to reveal any defect. It seems paradoxical. After all, each of these genes was the product of evolution and so presumably fulfilled some important task. It seemed counterintuitive that discarding these genes could have had so little consequence. And adding to the mystery, similar results have been found in other systems. Knockout mice have been produced that lack a protein that forms filaments; an enzyme that makes a crucial carbohydrate in the spaces between cells; a protein intimately involved in the generation of movement. And the result has been zero, null: there has been little if any obvious change.

Each case probably has to be considered on its own merits, and the reasons for a lack of detectable effect will be different for each. But Bé Weiringa has perhaps the most accessible system for analysis. Pathways of energy metabolism are so well understood that it is possible to understand how they function in the absence of particular enzymes. His answer to the puzzle, in broad terms, is that knockout mice compensate for defects by rerouting metabolic pathways. Other enzyme activities in the defective animals change in level so that they shoulder transformations normally performed by creatine kinase. In other words, he thinks that the mouse brain contains a network of complementary enzymatic pathways. When brain creatine kinase is deleted, other systems adjust their activities to restore the metabolic-energy supply. It is, of course, exactly what you would expect from the neural-network view of cell reactions: functions are shared among multiple units (that is, enzymes) so that the effects of damage to any individual unit or connection are mitigated. Graceful degradation indeed!

But wait a minute . . . isn't there something fishy here? If the enzyme creatine kinase can be removed without effect, why does the mouse bother to make it in the first place? Why does it make all the other genes that have been shown to be dispensable? Why doesn't the animal simply delete the redundant genes and save itself unnecessary expenditure of material and energy?

The answer is a biological one and comes from the realities of survival in the world. In every case that has been closely examined, gene knockouts do have *some* effects. They might be subtle and difficult to find under laboratory conditions, but the mutated mice are unquestionably defective. In the cruel killing fields of their natural environment, those mice would almost certainly die before their time. Wild animals

live on the brink; they are assailed from all sides by predators and parasites; they face starvation and disease daily. Any inherent weakness, such
as poor hearing or a less than optimal ability to remember places, will
make them relatively vulnerable under stress—losers in the game of life.
Even if these mice did manage by good fortune to survive and produce
a litter, their pups would inherit the deficiency. Sooner or later the lineage would disappear from the gene pool. The resilience revealed in
knockout experiments would not protect the species.

There's an echo here of the reality distortion of computer simulations.
The environment of a neural network—indeed any network designed by
humans on a computer—is far simpler than anything a living cell experiences. Neural networks can cope with large amounts of information,
supplied as written text, facial features, industrial processes, and so on.
They can recognize restricted sets of patterns better than we can. But
what they do is the tapping of a tin drum compared with the symphony
orchestra of natural environments. Cells have to respond appropriately
to conditions that change in myriad unpredictable ways. Consequently,
the networks of biochemical reactions that control cellular events—
everything from the copying of DNA and the synthesis of proteins to the
generation of energy from food and the production of cellular building
materials—are built to accommodate changes. Not the fixed stereotypical
step-changes of a training set presented to a neural network, but an overwhelming deluge of plurality, a barrage at all scales and in all channels.

I've argued in this chapter that small circuits of proteins have properties
in common with the computer-based neural networks. They can learn
to recognize patterns and abstract significant features of their environment. They can store this information in a manner that is resistant to
damage. Many other illustrations of a similar nature could be cited. The
Oxford physiologist Denis Noble describes the emergence of heart-
beats in his computer simulations of the heart. The periodic contractions of heart cells are due to cyclic changes in calcium ions, themselves

due to changes in the passage of ions across the heart cell membrane. But as Denis remarks, none of the proteins that transport different ions across the membrane is itself an oscillator. The periodic rise and fall of calcium ions—the steady beat of the heart—is the result of group activity: ion pumps and channels working together in the same cell.

Real protein networks are larger and more intricate than any neural network. Even the protein interactions within a single bacterium would—if we included all interactions—defy analysis. And if we consider the signals passing between the protein circuitry and the genetic circuitry of DNA and RNA, and between similar but nonidentical cells in a large plant or animal, we will indeed be at sea.

But perhaps you have seen enough of networks for a while. In the first chapter, you may recall, I raised the issue of cell awareness. Single cells such as bacteria, amoeba, and stentor display in their responses knowledge of their environment that informs all of their actions. It seems a reasonable guess that this knowledge is somehow written in protein molecules. Somehow the internal circuits of proteins, and hence the molecular structures of these proteins, have to encapsulate features of the outside world.

SEVEN

Cell Awareness

I am scribbling these notes seated on a hill in Scotland . . . one of those rare days on the West Coast when the clouds and mist roll away and the views have a crystalline clarity. Before me the Sound of Mull glints in the morning sun: wind-etched, steel-tempered blue. In the distance I see Ardnamurchan and the mainland, the Isles of Rhum and Egg, and farthest of all, the Cullins of Skye, like flat stage sets in shades of purple and gray. A cool wet wind blows in my face and whispers in my ears; a seagull flies across my line of sight, a sheep bleats. I smell the grass. I feel its slight dampness on my hands as I spread my arms to grasp the springy turf between my fingers. If I introspect further, I am astonished, not for the first time, by the clarity of the visual scene. It seems almost as though a hole has been scooped in my head. I sense the massive presence of the ground that, now I think of it, seems to press upward on my feet and back. Focusing, I feel the tidal sweep of air in and out; the slight expansion and contraction of my ribs as I breathe; the beating of my heart, still pounding from the climb; a slight feeling of emptiness from my stomach. I reflect that breakfast was several hours ago, and my thoughts swoop down to lunch yesterday in the pub at Dervaig. They wing back, following a habitual flight path, to the writing of this book.

This, as best I can describe it, is the experience of being alive, now. A confluence of sensory and mental stimuli centered on an apparent "I." It may be particularly vivid this morning because of my being in a new place and away from social interactions and computer screens. But otherwise it is completely familiar. There is an outside, whence comes a continuous stream of sensation. There is an inside, where the "I" reposes, collecting and, in a more or less deliberate manner, processing these sensations. My responses to these sights, smells, and physical pressures are edited and controlled by my thoughts: my sense of who I am, where I am, what I have been doing, and what I will (or should) do in the future.

Now what if my dog Skip could be here beside me? What would his sensations be? How much of my sense of identity and selfness would he share? I cannot tell: this is the Catch-22 of consciousness. Awareness of self is such a personal thing that no one else can share it, inaccessible to external analysis and scientific experimentation. However, after long observation of Skip and other dogs, I find it inconceivable that they would not share our experience of the world. I imagine a confluence of sensations, different in some respects from those I experience (more intense in the case of smells, for example). He would have felt the grass and been aware of my presence beside him (as I would be of him). He would have seen the same view, albeit from a different perspective, and would have learned already about his location—the pathway here down the rocky path. His emotions would have a similar physiology to my own, if more impulsive.

But Skip's muteness would make an essential difference. Not just because it renders him incapable of communicating his internal state to anyone else but also because it would leave that internal state so impoverished. Without language a creature lacks concepts. The distinction between self and not-self—even the separation between self and surroundings—will be vague at best. And what about our sense of the passage of time? This is inextricably enmeshed in language and especially dependent on metaphors of space (consider the origins of the word *passage* in the previous sentence). We mentally allocate times to positions in space, so that the farther away they are in time (past or future) the farther away they are portrayed in physical distance. We measure our lives by time lines and calendars. But wordless Skip has none of this. He is unable to share our sense of time: he has no past or future, just the here and now.

Next consider a mouse . . . perhaps there is one nearby crawling in that narrow crevice beside the rocky outcrop. It is much harder for me to empathize with such a tiny creature—not only because of its diminutive size but also because of my unfamiliarity with its needs and priorities. Wild animals, unlike dogs, have no need to react to human expressions and gestures. Their world is full of wild smells and tastes. A mouse must have knowledge of grass roots and stones, memories of past dangers, recollections of other mice that we do not share. But for all that, are they not built of flesh and bones like us? Certainly a mouse has a nervous system built on the same plan as ours (a fact much exploited by medical researchers). Given such similarities in anatomy, is it conceivable that this creature does not have a sense of place, a feeling that it is a separate organism distinct from its environment? Surely it will have internal states similar to what we term emotions. Mouse hormones are chemically similar to ours and have comparable effects on physiological functions of the heart, digestive system, and adrenal glands.

Now imagine that hidden beneath the rocky outcrop there is a beetle foraging in the damp recesses, potential food for the mouse. Here is a truly alien creature, a miniaturized robot encased in black armor. I have little chance of empathizing with such a creature or of knowing what feelings, if any, it experiences as it clambers over stones harvesting fragments of vegetation and fungus. Given the relative simplicity of its nervous system, the prospect that any insect could be capable of introspection or reflective thought seems impossible. Nor will it be capable of a great deal of learning: most of its actions will be of the hard-wired, reflexive variety, like the Sphex wasp mentioned in Chapter 1. But on the matter of sensations I am less sure. The exigencies of any self-propelled organism surely present common problems in the detection and selection of a path. A beetle will presumably experience the ground and sense where its body is—it must know where each of its six legs is in relation to its body. If its body loses nutritional reserves, the organism will experience hunger. A mature beetle might feel a desire to find another beetle of the same species but different sex to mate with, or a place to deliver offspring.

You can see where I am going. If I track an evolutionary staircase from humans to progressively lower organisms, there seems to be continuity. At each step I recognize structures (legs, eyes, mouth-parts, reproductive organs) and behaviors (escape responses, avoidance of pain, searching for food, seeking a mate) from the next-higher species. So if evolution has produced a commonality of structures, behaviors, and molecules, why should not the same be true of such associated subjective feelings as fear, pain, hunger, and lust? It is true that, in the case of feelings, we do not currently understand the cellular and molecular mechanisms involved. But is this reason to deny their existence?

You might tell me at this point that I have things upside down. Certainly any biologist can tell you that dogs, mice, and beetles are not in fact antecedents of humans in an evolutionary sense. Each has arisen by an independent lineage from a common ancestor. Most have gone through many more generation cycles than humans because they reproduce faster, so in that sense they are more highly evolved than we are. But this conclusion only makes my argument stronger. From a broad biological perspective we need to ask not "Which human sensations are also experienced by animals?" but rather "Which animal sensations are shared by humans?" Surely it is arrogance to declare that we alone experience emotional drives and internal sensations.

Arguments about the nature of sensations experienced by organisms other than humans have been going on for centuries. Julian Jaynes, in his 1976 *The Origin of Consciousness in the Breakdown of the Bicameral Mind,* offers an entertaining précis of the main positions. His own youthful quest for the roots of consciousness began with protozoa and the plant mimosa. He searched for evidence that these creatures could learn—regarding this as a test for awareness. But on mature completion of his odyssey he concluded that consciousness is not only an essentially human attribute but also one that appeared in historical times. In his view, the slaves who built the pyramids and the foot soldiers of the *Iliad* were not fully conscious in our modern sense.

At the other end of the spectrum, Herbert Jennings (see Chapter 1) remained uncommitted but was at least open to the possibility of sentience in single-celled microorganisms. His contemporary the German psychologist Wilhelm Wundt went further and articulated the view that it was "a highly probable hypothesis that the beginnings of the mental life date from as far back as the beginnings of life at large." The most convincing demonstration of mental life for Wundt was the display of voluntary actions in response to feeding or sexual impulses, both displayed by microorganisms.

There has been no resolution of these issues. The consensus of today's biologists is that conscious experience is the prerogative of higher animals. Most will assert if asked that consciousness requires a nervous system, and probably a well-developed social structure and even a level of language as well. But if I put to one side the thorny issue of consciousness, this still leaves a whole spectrum of possible interactions with the world. All organisms are inundated by stimuli and sensations every second of their existence. They have to assimilate this flood—rationalize, prioritize, store memories, make decisions. Because these organisms are connected by the lineage of evolution, it seems inescapable that they will use common strategies.

Let me start on solid ground. Take another aspect of conscious self that is better understood: one that illustrates the power of molecular analysis. What I have in mind is respiration. A mundane process, you might say, and yet an integral part of our sense of self. As with many other aspects of consciousness, we are not aware of our breathing until someone calls attention to it, or it suddenly stops. As Charles Darwin comments:

> A hundred muscles . . . are employed every time we breathe; yet we take in, or let out, our breath, without reflecting what a work is thereby performed: what an apparatus is laid in, of instruments for the service, and how many such contribute their assistance to the effect! Breathing with ease, is a blessing of every moment; yet, of all others, it is that which we possess with the least consciousness. A man in an asthma is the only man who knows how to estimate it.

From a mechanistic standpoint, breathing holds few mysteries for us today, thanks to our scientific culture. But this was not always the case. The word *respiration* has the same root as *spirit,* and most early historical cultures assigned a mystical significance to breath, using such terms as *ka, chi, thymos,* and *prana.* To someone from a nonscientifically based culture, breathing could indeed seem one of those miraculous processes, like the heartbeat and pulse: continual evidence of our vital state. Our life begins with our first breath and ends with our last. Our cycle of breathing (the durations of inspiration and exhalation) changes as we laugh, attend deeply to something, or feel surprise or sudden fright. Why should breathing be any less mysterious than other aspects of human existence, such as consciousness? But in contrast to consciousness, respiration now has a completely logical scientific explanation.

The study of respiration and the eventual elucidation of its mechanism are interwoven with the rise of modern science. In the eighteenth century, breathing was interpreted in terms of the phlogiston theory, first propounded by George Stahl. This is the idea that the heat and flame of burning materials result from the release of a substance called phlogiston. Plants were thought to absorb phlogiston from the air and incorporate it into their tissues, whence it could be released by burning. Animals that ate plants assimilated phlogiston into their own tissues, then released it during their breathing—a form of combustion. Joseph Priestley, the pragmatic Presbyterian minister from Manchester, who in 1774 codiscovered oxygen and showed it to be essential for the survival of a mouse in a jar, interpreted his results in terms of phlogiston. His name for oxygen was *dephlogistated air*—that is, not a single substance in its own right. A glowing taper in oxygen, the theory went, bursts into flame because it releases the stored phlogiston.

We understand the scientific basis of respiration today because of Antoine Lavoisier, a member of the French bourgeoisie who eventually lost his life on the guillotine. Lavoisier, a self-financed gentleman scientist, devoted his wealth and talents to addressing a wide range of fundamental issues in chemistry, including such ideas as the transmutation of elements and the existence of phlogiston. His most important experimental innovation was to make accurate measurements of the weights of substances before and after a chemical transformation. He thereby demonstrated that metals heated in air increase in mass owing to the incorporation of the new gas he later termed oxygen. Water was therefore not a single element, as the alchemists had long supposed, but itself a compound composed of oxygen and hydrogen—another of his newly coined names. Indeed, Lavoisier provided the basic nomenclature for chemicals still in use today. But his most important contribution was to declare and establish the principle of the conservation of mass: however a substance changes, its mass will remain the same so long as it is in a sealed container. Through his efforts, alchemy was transformed into the modern science of chemistry—a true Scientific Revolution.

Lavoisier's specific contribution to the understanding of respiration culminated in a series of experiments he performed with the eminent mathematician Pierre-Simon de Laplace. They already knew that respiration involved the consumption of oxygen and the production of carbon dioxide. So did a burning flame. But what was the relation between these two? In a paper published in 1783, Lavoisier and Laplace showed that the burning of charcoal and the respiration of a guinea pig produced the same amount of heat for the same quantity of carbon dioxide. In 1790 Lavoisier presented a summary of his findings, concluding that "the purpose of respiration . . . [is] a slow combustion of carbon and hydrogen similar in every way to that which takes place in a lamp or lighted candle." It was the first important step toward the realization that vital living processes are in fact chemical reactions.

There was still a long way to go; no less than two hundred years of research was needed to arrive at the modern view of respiration. Food molecules were burned not just in the lungs or liver, as first thought, but everywhere, in every cell of every tissue. These tissues obtained their oxygen from the blood circulating from the lungs to the distant tissue. And how does blood carry oxygen? Why, by holding it tightly to the abundant blood protein hemoglobin—bright red when combined with oxygen and dark red when not.

Researchers also worked out the sequence of reactions that take place in all cells whereby carbohydrates, protein, and fats are progressively oxidized to generate carbon dioxide and water. The terminal events in this series of oxidations occur in mitochondria, wormlike organelles that inhabit the cytoplasm of every animal and plant cell. Oxygen brought to the cell via hemoglobin in the blood diffuses across the cell membrane, crosses the cytoplasm, and then diffuses into mitochondria. There it interacts with hydrogen atoms plucked from molecules of sugars and fats. The offspring of this marriage of two gaseous elements, oxygen and hydrogen, is the transparent liquid we call water.

The final stage of combustion turns out to be electrical in nature. A chain of electron carriers in the mitochondrial membrane allows electrons to flow from molecules of food to oxygen, generating a continuous supply of electricity. The flow of electrons pumps protons (positively charged particles, also known as hydrogen ions) out of the mitochondria, causing a positive charge to accumulate on the outside. In 1961 an English biochemist, Peter Mitchell, proposed that the difference in charge, or electrical gradient, across the inner mitochondrial membrane could be used as a source of energy. His suggestion that the proton gradient could be used to drive the synthesis of the energy carrier ATP is now universally accepted.

Many single-celled organisms use atmospheric oxygen to drive energy production. They also perform what is described in the textbooks as respiration—although it appears at first sight to have little in common with the human activity of that name. An amoeba at the bottom of a pond has no direct contact with oxygen in the air. Its supply of the gaseous element comes from the limited amount that dissolves in water. Uptake into the cell itself is by passive diffusion: the small, uncharged oxygen molecules simply slip through interstices between the lipid molecules of the membrane. Carbon dioxide molecules produced at the end of metabolism exit by the same route. There is nothing comparable here to the Baroque labyrinths of our lungs or the tidal wash of air created by our chest muscles. Nothing equivalent to the rapid-transit system of our blood, crowded with carriers that pick up oxygen molecules at one stop and transport them within seconds to another stop at a distant location. It all seems very different from human respiration.

And yet . . . if you track the path of one oxygen molecule as it enters the amoeba, you will see the connection. For its eventual target is none other than a mitochondrion, the very same organelle that provides the final target of oxygen taken in by human lungs. These tiny membrane-enclosed organelles evolved specifically to perform the last stages of energy production from food. Just as in a human cell, or indeed the cell of any plant or animal, mitochondria in an amoeba have the capacity to react specifically with oxygen. There are proteins in a mitochondrion that carry an atom of iron or copper held in such a way that they control its oxidation: they determine when oxygen and iron or oxygen and copper combine.

Once oxygen is trapped in this way, it begins the cascade of electron shifts mentioned above. Electrons move downhill, in energetic terms—away from oxygen and toward hydrogen and carbon atoms. But instead of creating heat, as it would if it were allowed to occur by simple combustion, the process in a mitochondrion is fed through a series of biochemical intermediates like water in a cascade. Moving from carrier to carrier, the energy is captured in chemical form. Through the good offices of carrier molecules, honed by evolution to perform just this task, high-energy electrons drive protons from one side of the membrane to the other. The gradient thereby created is a source of stored energy—a hydrogen fuel, no less. Most important, the proton gradient drives miniature rotating protein machines embedded in the membrane. As these nanogenerators turn, they make ATP molecules. ATP then diffuses out of the mitochondrion and into the cell, where it drives a myriad of energy-requiring processes.

So the terminal events of respiration in an amoeba are closely similar to those in a human cell. Indeed, they have a family resemblance. Mitochondria from an amoeba not only look the same as mitochondria from virtually any other organism, including humans, they also perform the same sets of reactions and use proteins that are virtually identical.

Cytochrome c is one of the proteins that carry electrons in the membrane of a mitochondrion. This ancient protein evolved more than 1.5 billion years ago—before the divergence of plants and animals. Its function has been conserved throughout this vast period. Properties such as color and electron affinity are virtually indistinguishable for cytochrome c's in all mitochondria, whether from plants, animals, or protozoa. In test-tube experiments, the cytochrome c of any one species, such as wheat, can take the place of the protein from any other species, such as human.

But the most telling evidence comes from the sequences of amino acids of cytochromes from eighty different species. The striking finding is that out of 104 amino acids in the protein, 26 are the same in all species. In other words, 26 amino acids have been retained in the same position of the same protein generation after generation, for more than one and a half billion years of evolution. Why these particular amino acids have remained constant becomes plain when one examines the structure of the molecule. Some are linked to the atom of iron that interacts with oxygen. Others make up the site that recognizes and binds to other proteins in the membrane. In short, the conformation of the protein—that is, the three-dimensional arrangement of its polypeptide backbone and the location of most of its atoms—is essentially the same in a human cell and in an amoeba.

A similar argument can be made for other fundamental processes. Take the way we walk and run around, powered by the muscular activities of our legs and arms. A nonbiologist might suppose that this has absolutely nothing to do with the slow creeping motion of a microscopic amoeba over the bottom of a pond. But a biologist knows otherwise. The motion of eukaryotic cells over surfaces is built around the same two proteins, actin and myosin, that drive muscular contractions in our body. The proteins are unmistakably the same, in shape and chemical performance and amino acid sequence. They produce movement in the same way at the molecular level, with myosin progressing step by step along actin filaments, driven by the energy molecule ATP.

Actin, in particular, is highly conserved. Actins from different species as far apart as plants, yeasts, and mammals have similar amino acid sequences: actin from a soil amoeba differs from mammalian actin in only 15 out of 375 amino acids. It is therefore no surprise that these proteins also function at the molecular level in the same way. Once again, I argue that the same function in a human and an amoeba is based on an essentially identical molecular interaction. The evidence from sequence analysis shows that these two forms of movement are descended from a common ancestral organism: presumably this ancestor was single-celled and in that regard closer to the present-day amoeba. The movements of a leopard or a lynx—vigorous, violent, and at a much larger scale—can be seen as the result of amplification in size and progressive specialization in performance. The actin and myosin molecules in these animals are installed in large numbers of relatively huge muscle cells, dedicated to the production of mechanical contraction.

This conclusion leads me to ask whether a similar lineage could be traced for what we might call knowledge of the environment. All living organisms—at least all of those that lead a motile free-ranging existence—can detect heat, light, and mechanical vibrations. They locate food and avoid noxious stimuli. Many *seem* at least to display pain, fatigue, hunger, lust. Yes, there is an enormous chasm between the manifestation of these states in an amoeba and in higher animals such as humans. But because knowledge of the world is such a fundamental part of life, essential equipment for an organism, it seems to me that there must also have been continuity, an unbroken chain of past organisms that displayed these or similar actions. It seems reasonable to suppose that, as for respiration and movements, evolution could have preserved essential molecular interactions able to capture features of the world. Properties that, once they appeared, were too useful and powerful to ever discard.

My argument is that living cells have an intrinsic sensitivity to their environment—a reflexivity, a capacity to detect and record salient features of their surroundings—that is essential for their survival. I believe these features to be deeply woven into the molecular fabric of living cells. A primitive awareness of the environment was an essential ingredient in the origins of life: it was preserved and expanded in increasingly elaborate and prolific forms in the subsequent explosion of living organisms during evolution. This intrinsic capacity has been amplified and ramified in a thousand different ways. It is the molecular substrate of our knowledge of the world, including our sense of self— the seed corn of consciousness.

This does not mean, I should emphasize, that they are the same thing. I do not believe that amoebae possess sentience any more than they possess legs or lungs. Consciousness in particular may well be a comparatively recent invention, appearing in some advanced animal by a rare conjunction of features, essentially sui generis, without significant precursor. Human language is one attribute that could fit this description, since despite impressive performances by chimpanzees, dogs, and parrots in word recognition, no other animal can communicate as we do. It would be futile to look for a primitive language center in the nervous system of a fruitfly or a nematode. But I nevertheless expect to find in these animals knowledge of the world and their place in it.

EIGHT

Molecular Morphing

If responsiveness to the environment is an essential ingredient of all living forms, then logically it should have appeared at the same time as the first organisms. There are many debates about the origins of life. Even the location is uncertain, with some reputable scientists even suggesting that it may have occurred not on Earth but elsewhere in the solar system, or farther away. The precise sequence of chemical events that led to the creation of the first self-replicating entity may never be known, because no relics remain. But from our knowledge of present-day organisms, it is certainly possible to make educated guesses about the likely milestones on the way. Could they chart the origins of self-knowledge?

The traditional place to start is with a broth of simple organic molecules. In now classic experiments performed at the University of Chicago in 1953, Stanley Miller and Harold Urey passed electrical discharges through a glass flask containing a gaseous methane, ammonia, hydrogen, and water. After one week, they found carbon atoms of the methane incorporated into simple organic molecules, including amino acids. Similar experiments later revealed that most of the small molecular building blocks of life could be produced under conditions resembling (somewhat) those on the cooling earth. These include lipids similar to those in cell membranes, sugars used by present-day cells for energy production, amino acids such as those in proteins, and the nucleotide bases that form the building blocks of DNA and RNA. These processes are often extremely slow in the laboratory, but this seems immaterial when the Earth had millions of years available. Moreover, there were no bacteria or other organisms at this stage to consume any molecules made—nothing to prevent organic chemicals accumulating in high concentrations.

Simple substances such as amino acids and nucleotides can react spontaneously to form large chainlike molecules (polymers). Amino acids of twenty kinds join to build proteins; four kinds of bases join to make RNA and DNA. Biologists speculate why this particular set of twenty amino acids and four nucleotides were chosen out of hundreds of chemically similar molecules; we may never know the real answer. But that they recur in all existing life forms testifies to their common origins.

Of all the complex molecules found in living cells, ribonucleic acid, RNA, was probably most important for the origins of life. RNA is a linear polymer formed from chains of nucleotide bases. It was found in the 1950s to be a molecular copy of regions of DNA—an essential go-between that transports the sequence of a gene to the machinery of protein synthesis. But as time has passed, so the number and variety of functions of RNA molecules in cells have increased. RNA molecules are able to fold in distinct ways and can provide frameworks for structures in the cell, sometimes coated with proteins like flesh on bones. RNA molecules can act like enzymes and accelerate reactions; they can adopt different molecular shapes, like allosteric proteins, and share many of their computational properties. A world of small RNA molecules has been uncovered that silence or modulate the expression of genes, especially in higher animals.

One of the first to understand the pivotal role of RNA in the origins of life was the founder of molecular biology, Francis Crick. In a preface to a collected volume dedicated to the RNA world, James Watson, who with Crick discovered the structure of DNA in 1953, recounts: "The time had come to ask how the DNA → RNA → protein flow of information had ever got started. Here Francis was again far ahead of his time. In 1968 he argued that RNA must have been the first genetic molecule, further suggesting that RNA, besides acting as a template, might also act as an enzyme and, in doing so, catalyze its own self-replication." The full implications of this idea, however, were not apparent until much later.

The term *RNA world* itself was first used in a 1986 letter to *Nature* by the Harvard molecular biologist Walter Gilbert. He observed that if large populations of self-replicating RNA molecules had existed, they would have been subject to evolution. Changes in the sequences of the RNA molecules, caused by errors in copying (mutations) and perhaps by wholesale exchanges of pieces of RNA polymer, would have led to new molecules with previously untested properties. Natural selection, he suggested, would have acted like a sieve to filter out variants that could not compete. Those that survived would be progressively improved with an ever-widening repertoire of enzymatic activities. Eventually, RNA molecules would begin to synthesize proteins from the pool of readily available amino acids. Because of their richer chemistry (based on twenty assorted amino acids rather than four rather similar nucleotide bases), proteins make better enzymes than their RNA counterparts. Later still, cells would "invent" the closely related molecule DNA and use it as a long-term storage of hereditary information. RNA would then be relegated to the role of a go-between, carrying information from DNA to protein.

At the heart of the notion of an RNA world is a stunningly simple chemical rule. Known as base pairing, this rule states that each base of RNA or DNA binds preferentially to one of the other bases. Because there are only four bases, this reduces to just two interactions. For RNA the four bases are U (uracil), C (cytosine), G (guanine), and A (adenine), and the two pairs are A plus U and G plus C. For DNA the base T (thymine) stands in place of U, so that the four bases are T, C, G, and A, and the two pairs are G/C and A/T. These simple pairings underlie life on Earth and provide the basis for inheritance and evolution. They also, from our present standpoint, provide a crucial clue to the essential difference between biological molecules and those made by nonliving processes.

The first consequence of base pairing is that two molecules of RNA or DNA will selectively pair, or hybridize, if their sequences of bases match. That is, if we imagine moving along a molecule, each base in one strand will pair with the base in the other strand. If this condition exists, then all the many base pairs will add to the strength of the interaction, and the two molecular strands will hold together, as in a clothing zipper. Hybridization is a fundamental property exploited by molecular biologists to select and manipulate DNA molecules. It is also important in many events in cells, as when chromosomes come together and exchange portions of DNA.

The second consequence of base pairing is that it empowers one RNA or DNA molecule to direct the synthesis of a second—effectively to make a copy of itself. Suppose you have a mixture of free nucleotide bases U, C, G, and A, floating free in solution. Suppose also that a suitable enzyme is available to string these bases together to make an RNA molecule. Now, with nothing to guide the reaction, it will simply make a random mixture of molecules. It will make chains such as GGAUCGG and AGAAAUCCCAA and so on, with various lengths and no particular order. If, however, if you also add to the mixture RNA strands with a certain sequence, let's say AGGGUGGGAGGGUGGG, then these can guide the synthesis. Under the right conditions, the added polymer will act as a template that directs the sequence of newly added bases so that it makes a copy of itself.

The details are somewhat complicated because each template directs not an exact copy of itself but a complementary copy—one that forms base pairs along its length. The first product is a pair of strands hybridized to each other. But both strands contain exactly the same information. Pull them apart and each can itself act as a template and direct further chains. In this way the original sequence AGGGUGGGAGGGUGGG will multiply and spread through the mixture of bases.

As Watson and Crick realized in 1953, the ability of DNA to generate strands of complementary sequence explains how hereditary information is passed from an organism to its progeny. The DNA of a cell is naturally two-stranded, consisting of two intertwined strands of complementary sequence. As a cell divides, a protein machine moves along the DNA, pulling apart its strands and making a complementary copy of each, in the manner just described. In the wake of the copying machine, each newly formed strand folds back and hybridizes with one of the two older strands. In this way, two double-stranded DNA molecules are created, each with the same sequences of base pairs. These two molecules, each carrying the same genetic information, move to different daughter cells after division.

The situation for RNA is less obvious, since this more flexible polymer is not always two-stranded. Nevertheless, RNA molecules can and do replicate in living cells. They also do so in test tubes when supplied with the correct enzymes. Similarly, it is believed, systems of self-copying RNA molecules became established on Earth during the origins of life. The first RNA molecules might have formed in several ways—for example, by the heating of dry organic compounds or by the catalytic activity of high concentrations of inorganic polyphosphates. These would initially have formed polymers of variable length and random sequence—that is, the nucleotide base added at any point would have been governed largely by chance. However, once short lengths of RNA had been created, then they could have been copied—used as templates to direct the synthesis of other similar molecules. Other RNA molecules, acting as enzymes in the manner described shortly, would certainly have been a factor.

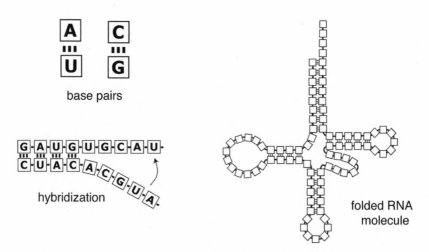

FIGURE 8.1. Base pairing in RNA. Nucleotide bases A and U recognize each other by forming weak chemical bonds, as do C and G. Chains of bases bind preferentially to each other if all the bases can form pairs—a process called RNA hybridization. Many RNA molecules fold into complicated shapes owing to base pairing between different parts of their long chains.

Base pairing is not only the way RNA (and DNA) molecules make copies of themselves. It also explains how a system of molecules can undergo something very like natural evolution. We know this because of laboratory experiments using RNA polymerase—the enzyme that catalyzes the replication of RNA molecules. A few molecules of a particular RNA, placed in a bath with nucleotides and RNA polymerase, will produce, like a pair of mating rabbits in a field of carrots, a million offspring. Most of these new RNA molecules will be identical to the original, so that if a fresh sample is taken and the process repeated, exactly the same thing will happen. However, the newly made population is never *exactly* the same as before. There are always small changes in the sequence—the wrong base inserted here or there, or perhaps one base missing or an extra one added.

The frequency of these mistakes (known, in other contexts, as mutations) can be increased experimentally. After a series of steps, RNA molecules begin to drift in sequence, and this is where differences in rate of replication arise. Some sequences will replicate faster or more accurately than others. One factor is the length of the molecule: within limits, shorter RNA molecules will be copied faster, so that more of them will be made. Or it could be that some RNAs fold so tightly that the RNA polymerase has trouble copying them. Another trick played by researchers is to introduce a second enzyme into the replicating mixture, one that degrades RNA to its constituent nucleotides. Any RNA molecules that can resist enzymatic breakdown then become selected, and eventually the entire population becomes resistant to the degrading enzyme.

Mentioning the folding of RNA brings us to the fourth and last consequence of base pairing. RNA molecules are not just strings of symbols like barcodes, carrying information and nothing else. They have chemical personalities that affect what they do. The sequence of nucleotide bases of a RNA molecule not only specifies the information that it carries. It also determines its relevant physical properties, notably, how it folds in solution. The reason is not hard to understand when we realize that base pairing works not only between different molecules but within a single molecule. For example, suppose an

RNA molecule has the sequence AAAA in one region and UUUU in another region. Continual motion generated by heat energy will cause the long molecule to flex and bend and by chance bring the two regions together. There they will stick, held together by complementary base pairs: each A will bind to a U. If the RNA sequence has multiple sticky regions of this kind, they will act like sutures or staples holding different parts together. The RNA molecule will fold back on itself in loops, forming a more compact and potentially complicated structure. Hence the RNA molecule will adopt a distinctive shape that is determined by the sequence of its nucleotides.

In 1989 the Nobel Prize in chemistry was awarded to Sidney Altman and Thomas Cech. They had discovered that certain kinds of naturally occurring RNA molecules spontaneously excise—cut themselves out of—longer RNA molecules. That is, they catalyze the fission of bonds between specific nucleotides without the need for a protein. This novel finding initiated a hunt for catalytic RNAs with other functions. Within a few years, other self-cleaving or self-splicing RNAs were identified that make single cuts in RNA molecules and rejoin the cut ends. Some of these catalytically active "ribozymes" occur, for example, in ribosomes—the large particles built from protein and RNA molecules that are used to make new protein molecules. Ribosomes move along a messenger RNA molecule, like a sewing machine, stitching amino acids into a growing length of polypeptide chain. For many it years it was believed that proteins performed all of the important steps of this miniature machine. The RNA molecules were viewed simply as a passive framework. Quite the opposite: we now know that the actual catalytic site for many of these reactions is RNA.

Advocates of the RNA world are delighted with these results. If RNA molecules can cut and splice each other, then, they suggest, there is little reason why they should not be able to catalyze their own replication. Indeed, RNA catalysts that link pairs of molecules have been created under laboratory conditions. Ribozymes have been made that can interact with amino acids or with lipids. There has also been technological spin-off. Biotechnology companies now invest heavily in the large-scale production of ribozymes using techniques that, in a crude sense, mimic natural evolution. Starting with huge collections of different RNA molecules, they put them through repeated rounds of selection, increases in number, and changes (ersatz mutation) until the population eventually fulfills some criterion.

"Nothing in biology makes sense," the geneticist Theodosius Dobzhansky famously wrote, "except in the light of evolution." And this is why ribozymes are so important. Each RNA molecule, we see, possesses two vital characteristics. One is information, encoded in its nucleotide sequence and passed on by its replication. The other is biological function, determined by its unique folded structure. These features are the two essential ingredients for evolution. In a replicating population of RNA molecules, particular species will be copied faster or more effectively and hence be selected. Spontaneous changes in nucleotide sequence will alter molecular folds and hence influence molecular function, whether in replication or in some catalytic activity. A closely parallel sequence of events is thought to underpin the evolution of present-day living organisms, except that it entails a constellation of molecules of many kinds. The nucleotide sequence of an RNA molecule is analogous to the hereditary information, or genotype, of an organism—it is equivalent to the code that is carried, in present-day organisms, by DNA. On the other hand, the folded three-dimensional structure of an RNA molecule is analogous to the phenotype—the physical and biochemical properties of an organism. The phenotype of a plant or animal is a collective feature of its many different molecules, but it depends above all on proteins.

Speculative accounts of the evolution of systems of RNA molecules eventually run into a conceptual barrier. The problem is as follows. We can picture that, through the process of mutation, one kind of RNA molecule might find how to better reproduce itself. But if this molecule is simply released into a uniform sea of other RNA molecules, then it will share its discovery. RNA molecules with sequences that do not have special properties will be carried along like social parasites, benefiting from the hard work of others. Similar arguments hold, with even more force, when sets of molecules work together in a cooperative fashion. Say you had, in your evolving soup, a dozen species of RNA molecules that had found how to cooperate to catalyze formation of nucleotide bases, the raw material needed to make RNA. Without some form of segregation, these talented individuals would become separated and lose their cooperative abilities. Worse: any raw material they did make would diffuse freely through the population and provide food for all the other RNA molecules—including their competitors.

From the standpoint of evolution, this is exactly what we do not want. Natural selection works because organisms better at producing offspring multiply at the expense of their close relatives. So our population of RNA molecules needs something equivalent to an organism— some way to partition sets or cohorts of molecules. Cohorts must be sufficiently isolated and distinct that they can keep to themselves, self-ishly, any improvements they achieve.

Which brings us to membranes. All present-day cells are enclosed in thin sheets of lipid and protein. These membranes segregate the molecules responsible for the replication of the cell (as well as its growth, its energy production, and so on) from everything outside, including other cells. The utility of this arrangement is inescapably powerful. It seems unavoidable that at some stage in the RNA world, systems of replicating molecules must have become enclosed in spherical containers composed in large part of lipid.

Although this seems at first sight an almost impossible task, there are ways it could have happened. Lipid molecules have long oily tails made of hydrocarbon and a water-soluble head group. Because of this distinctive chemistry, lipids spontaneously form thin sheets quite similar to those in cell membranes—all you need to do is shake a drop of lipid with water. Lipid sheets are built like molecular sandwiches, with hydrocarbon tails on the inside and water-loving groups on the outside. Moreover, they tend to fold back on themselves into closed structures, typically spherical or tubular vesicles. Soluble ions or molecules in solution at the time of vesicle's formation naturally become trapped within and sealed off from the surroundings.

So it seems possible that a population of replicating RNA molecules, plus other organic molecules, at some time in the distant past became enclosed in lipid vesicles. We can only speculate how this may have happened. It may have helped if some of the reactions took place on surfaces rather than in solution. Certain types of mineral have interesting catalytic properties owing to arrays of negative charges on their surface. Small particles of montmorillonite, a common silicate component of clay, are known to bind both RNA molecules and lipids. They not only can promote formation of small RNA molecules but also generate lipid vesicles with enclosed RNA. Addition of fresh lipid molecules enables the vesicles to grow—they can even divide without dilution of their contents if the vesicles are physically forced to pass through small pores. It is a long way from a cell, but perhaps a start.

A plausible scenario would begin with the generation of lipid vesicles enclosing sets of RNA molecules. The latter should be capable not only of their own replication but also of promoting the incorporation of additional lipid molecules. As the vesicles grow larger they would also become more fragile, prone to break if their surroundings were shaken or agitated. Smaller vesicles made in this fashion might grow and fragment repeatedly in a crude imitation of cell division. Any vesicle that managed to achieve this much, however slowly and imperfectly, would have an insuperable advantage over all of its neighbors. It would be poised to take over the world—literally.

Many other hurdles would have to be overcome before such a primordial system could give rise to what we would acknowledge as a true cell—complete with DNA, protein, and membranes. Protein synthesis is a major challenge because it requires transfer of information between sequences of nucleotide bases (in RNA) and sequences of amino acids (in proteins). Other transitions in early evolution must have included the emergence of DNA as the primary storage of hereditary information, the acquisition of filamentous protein structures, and (for eukaryotic cells) the development of internal membranes and organelles. We are a long way from understanding how these events occurred, although many plausible hypotheses can be found.

Imagine Planet Earth three and a half billion years ago. A flat and monotonous terrain, pockmarked with irregularly shaped pools of water, extends to the horizon. Heavy rain-filled clouds brood overhead; the darkness is broken here and there by shafts of light. Distant pools flash like silverfish against the calico pattern of browns and grays. Like present-day tundra, but hot, humid, and threatening: lightning flashes, the earth shakes underfoot. There is no oxygen or obvious signs of habitation. But now look closely at the margin of this pond. See the oily rim? The water here seems opaque and muddy . . . except that this "mud" is primitive life. Countless protocells are here, sticking to the rocky substratum or stirring slowly in the tepid water. These membrane-encapsulated chemical packages have passed evolution's qualifying exam: they can grow and make more of themselves; they can divide. Only slowly and imperfectly, it is true. They take weeks to achieve what an *E. coli* bacterium does today in twenty minutes. Their division generates disastrous errors. Look: here are dwarf cells, giant cells, cells distorted by warts and protrusions: a freak show!

But these cells have what it takes to survive. Each tiny sac encloses a complete, functioning system of molecules capable of self-replication. DNA molecules multiply and portions are copied into RNA; chains of amino acids are linked in sequences specified by RNA. Some primitive proteins function as pumps to absorb nutrients; others act as enzymes to catalyze the breakdown of sugars (without the use of oxygen, as yet); others produce chemical energy in the form of phosphate-rich ATP. Using this stored energy, the protocells drive other reactions, including the manufacture of novel and increasingly large molecules.

And they already compete. Here, in this pond, a vicious struggle for survival is being fought. Famine, pestilence, and warfare already stalk the Earth, on a microscopic scale. Reserves of organic material are depleted; cells fight for food. Variant forms of life appear and scrabble to the top of the heap, enjoying a brief supremacy before being deposed in their turn. Some cells become specialized at harvesting scarce provisions, better able to use what is there. Other cells acquire the trick of floating or crawling to new locations more favorable for growth. Yet other cells learn the arts of war—how to kill and to cannibalize. Wave after wave of aspirants appear and are tested on the same killing fields. Most of these miniature warriors perish and leave no trace; a few survive long enough to deposit a genetic legacy.

Suppose we could travel back in time and select one primordial cell, an ancestor of present-day bacteria. Now imagine that we follow the history of this cell and its descendants as a time-lapse movie, speeded up so that a million years pass in a minute. What would we see? In this accelerated time frame, cell divisions pass in a flicker, a microsecond. The cell in the center of our view seems to shimmer as it repeatedly grows and divides, grows and divides. At each division, of course, two daughter cells are created, so entire populations might arise from our initial cell. But in our movie we ignore them by focusing always on one lineage. Changes in food supply, depletion of reserves, release of novel compounds by surrounding organisms, all these occur abruptly, within seconds. Environmental fluctuations occur in a flash: diurnal changes in temperature, longer seasonal changes. Now and then there are catastrophic events: hurricanes tear the surface, volcanoes erupt and the sun darkens. With each change in the environment, our single cell responds, reacts, changes. With each new constraint it becomes better equipped and more sophisticated.

And here's the point: it *looks* to our foreshortened, time-compressed view as though the cell is warned in advance of every change, as though it anticipated each emergency and already knew how to deal with it. But this is an illusion, a consequence of our method of tracking. Remember that we are following just one lineage—one that not only survives in this generation but also is an ancestor of a line of successful cells. What actually happens is that the cell continually undergoes random changes in its DNA. Mutations, duplications, and rearrangements of its genetic material alter its molecular composition. These change its form, composition, and what it does. They affect its ability to collect food molecules, digest them, make new substances, and compete with other cells. Harmful mutations, by far the most abundant, pass in the flickering of an eye. We do not see them because we have chosen not to follow the fate of daughters that die.

Mutations would have been much more prevalent at early stages of evolution: the machinery of copying of RNA or DNA molecules less effective, with many more errors in duplication. Moreover, the efficiency of repair, essential for the survival of present-day organisms, would also have been rudimentary. Another process that would have contributed to the rapid changes in genetic material is the wholesale exchange of large pieces of genetic material between cells. This can occur though simple contact or by the agency of an independent package of RNA or DNA in the shape of a virus. Whole genetic elements could spread though a population, introducing genes that carry particular features: once again, the most effective and more rapidly propagating would be those that show long-term survival.

In our brief lives we catch so little of the vastness of evolution that living species seem to us defined and immutable. But taking a long view and foreshortened perspective, we will see that plants and animals continually morph—seethe with change as new genes and new proteins are continually created and then lost. The linear sequence of base pairs in their DNA continually transforms as new bases come and go, segments disappear or are duplicated, and lengths of DNA move wholesale to new sites. All of these changes are triggered (or seem to us to be triggered because of our history glasses) by external events. Climate changes, the appearance of new chemical toxins in the environment, the emergence of novel viruses—each event presents a new challenge and a new set of selective pressures. A flurry of changes in the DNA then leads to novel arrangements of proteins and new phenotypic features.

Nor is mutation the only source of variation. The ability to adapt to environmental changes is far too important to be left to a single mechanism. Mutations may be the raw material of change, but (switching metaphors in midstream) good mutations are as rare as hen's teeth. It takes many generations before advantageous changes in the DNA can be installed throughout a population. This is far too slow to tackle many fluctuations that occur under natural conditions.

No surprise then, that cells discovered faster and more flexible ways to change their chemistry. Closely related to mutations are the DNA duplications, inversions, and other rearrangements that shuffle the genetic information. Processes of this kind can also create novel sequences, but their most telling effect is to put existing genes and sequences together in novel ways. A ubiquitous source of this shuffling in higher organisms is sexual reproduction. When two closely related sets of genes are brought together in the same cell, as occurs during the fertilization of an egg, new combinations of properties are created. The most important force driving this mechanism, experts in evolution tell us, is the continual war against parasites. Like the Red Queen in *Alice in Wonderland,* organisms have to run ever faster just to stand still. But it is also an engine for diversification leading to a continual expansion of an organism's repertoire.

Another source of variation is RNA splicing. A typical gene in a higher organism is not in fact a single contiguous stretch of DNA but a series of separate segments. Before the gene can be used to make protein, the different segments must be put together, "spliced," to make a single long RNA molecule. Although it seems wasteful, this curious procedure is thought to have significant advantages in flexibility. During evolution, for example, RNA splicing enables different parts of a large protein to assemble in different combinations. Parts of proteins that serve a distinct function, such as binding to a particular receptor or to the nuclear membrane, can be built like Lego bricks into different structures. From our time-warped, morphing perspective, protein segments are like single notes on a piano that are used repeatedly in many different chords.

Gene splicing also provides a major source of variation within a single animal. In this case, the RNA splicing machinery produces different outputs in different places—making one form of the protein in the liver, for example, and another in the brain. Even in a single tissue you can find many different forms. Many of the proteins in mammalian muscle, for example, exist in a large number of variants. Troponin T, involved in the calcium sensitivity of muscle, is made in more than eighty different forms, owing to alternative RNA splicing. Some are characteristic of the type of muscle, others change with exercise, and so on. There are probably as many different cellular blends of troponin T as there are Scotch whiskies.

The nervous system seems a hive of gene-splicing activity. For example, humans have just three genes for the protein neurexin, found on the surfaces of nerve cells. But splicing of the RNA produced from neurexin genes creates thousands of slightly different proteins. These are made in distinct combinations by different types of nerve cells. The gene for another protein on the surface of growing nerve cells, Dscam, has almost one hundred segments and can generate thirty-eight thousand possible variants. There is evidence in fruitflies that this plethora of Dscam molecules exerts a controlling influence on the connections between nerve cells. If the number of variants is reduced experimentally, then a major disruption of neural circuits in the fruitfly's nervous system ensues.

The old dictum "one gene, one enzyme" propounded by the Caltech scientists George Beadle and Edward Tatum in 1940 is clearly not true in detail. One gene can make more than one protein, sometimes many more.

Our description of the origins of life is the product of imagination. It has to be—we cannot go back in time to witness the crucial molecular events in the path of evolution. All we can do is to construct plausible scenarios for what might have happened based on processes that we know do happen now. And fortunately for our story, a vivid demonstration of molecular morphing actually exists in present-day organisms. It can be found in the immune system of humans and other mammals.

We live in a world of bacteria, fungi, viruses, and parasites, all eager to invade our nutritionally rich flesh, and we would be dead were it not for our immune system. This bodywide system of interacting white blood cells has a truly amazing capacity to recognize and destroy pathogens. Its importance to health is seen when it becomes disabled, as in the terminal stages of HIV/AIDS, or when it turns, inappropriately and aggressively, against its owner, in such diseases as rheumatoid arthritis or multiple sclerosis.

The immune system is one of the most elaborate cellular systems we know, and depends on the proliferation and interactions of many kinds of white blood cells—lymphocytes, neutrophils, macrophages, and so on. Lymphocytes themselves are made in myriad different forms (perhaps 100,000 kinds). Each is specialized to recognize, in the sense of binding specifically, a particular small molecule or part of a large molecule. A lymphocyte that encounters an invading microorganism or a molecule that it recognizes as foreign—termed an antigen—starts to multiply furiously. Its progeny recruit other white cells, and the entire army then launches a coordinated attack on the invading organism, like a swarm of angry bees. Target cells are killed—literally eaten—by the white blood cells. In another distant echo of our antecedents, white blood cells called macrophages engulf cell debris and even whole cells such as bacteria, as in the feeding of amoebae.

When a lymphocyte recognizes a foreign molecule, it does more than just divide. The weapons of defense are not only amplified but also improved and made more deadly. As they divide in the presence of an antigen, lymphocytes continually refine the molecules responsible for binding so that they bind more tightly. Eventually, the products of this selective multiplication mature into cells that release proteins, antibodies, into the bloodstream. Antibodies stick so tightly to their antigen that they render it harmless and trigger its destruction. Special cohorts of mature cells also create an immunological reservoir or memory, kept in the body prepared to respond to a second infection by the same organism.

As well as antibodies, secreted in soluble form into the bloodstream, other closely related proteins are carried on the surfaces of lymphocytes, attached to their membranes. The entire family of proteins, termed immunoglobulins, has an unparalleled ability to recognize foreign antigens. It forms the molecular basis of the body's ability to distinguish self from non-self—indigenous molecules from alien invaders. The discriminatory ability of immunoglobulins depends on molecular binding and employs the familiar principles of complementary shape fitting and weak bonding. Immunoglobulins are unique, however, in the enormous variety of slightly different forms they adopt, and the huge number of other molecules that they can collectively recognize.

The field of immunology is one of the most complicated in biology—full of special names for a plethora of special cell types and molecules. To keep things simple, I will here concentrate just on soluble antibodies. In molecular terms these are Y-shaped proteins with two identical antigen-binding sites, one at the tip of each arm. Their binary character is important, for it allows them to act as molecular cross-linkers. They bind tightly to soluble toxins such as those produced by bacteria, creating small particles that are recognized and engulfed by macrophages. Similarly, antibodies coat the entire surface of invading bacteria, not only rendering them harmless but also marking them for the attention of macrophages.

Individual antibodies are highly specific and bind tightly to just one or perhaps a few different antigens. This is why if we acquire immunity to one disease, such as mumps, we do not automatically become resistant to another disease, such as diphtheria. Researchers in laboratories routinely produce antibodies by injecting mice or rabbits with substances of interest. Bacteria and their products are particularly effective. But with persistence and skill virtually any substance can provoke specific antibodies, provided that it is not already present in the body. This applies even to novel chemicals that no rabbit could ever have met in the past. Clearly, therefore, we are not dealing here with a racial memory of a previous encounter. Rather, it seems that evolution has learned how to create versatile proteins that in a sense mold their shape to any foreign molecule. The number of different specific antibodies that a person can make is astronomical. Some estimates place it at around ten million trillion.

There is a problem here. How can our body make a virtually unlimited number of different antibodies when we carry little more than twenty-five thousand genes in our DNA? The answer is that the genes encoding antibodies undergo unique modifications—molecular morphing, if you will—as lymphocytes divide and mature in the body. In fact, what we inherit from our parents are not antibody genes as such but families of gene segments. Each family of segments encodes alternative versions of parts of the antibody molecules, distributed along the DNA like a linear patchwork. Before an antibody can be made, segments have to be cut out, rearranged, and recombined, a process that occurs only in developing lymphocytes. Not only are different segments cut out and put together in random combinations, but the splicing itself also generates additional diversity through the introduction of random changes in the DNA.

The first antibody molecules are produced in the body speculatively, as it were: many of them have not yet encountered (and may not ever encounter) a matching antigen. The reason for this strategy is that parasites and other invading organisms are forever changing their molecular surfaces, in an attempt to gain entry into the body. But the first meaningful encounter between one of these speculative lymphocytes and a foreign antigen sounds the alarm. Now the immune system knows it is under attack and starts to mount its defenses.

Any lymphocyte carrying an antibody that by lucky happenstance recognizes the foreigner is stimulated to divide. As this family of cells grows, the antibodies they make are in a sense shaped to the invading antigen. This happens not by the antigen interacting directly with DNA but by a kind of natural selection in miniature. A special mechanism in the lymphocytes introduces single mutations, exchanging one base for another into selected regions of the antibody gene. Substitutions occur at a rate of about one mutation per antibody gene per cell division; this is more than a million times greater than the ordinary mutation rate for a gene. Any cells that, through this random process, come to make antibodies that bind more tightly to the antigen are encouraged to divide rapidly. They multiply at the expense of their sister lymphocytes. Some of this new population becomes antibody-secreting cells in the blood. Others become long-lived memory cells that can respond quickly and forcefully to any future encounters with the same invading organism.

The immune system shows us that living organisms can shuffle their giant molecules like card players. Responding to an infection, they rearrange their DNA, within days or even hours—cobbling together genetic segments in new combinations. Using tricks, they focus random changes in particular regions of the genome at a million times the normal rate, submitting their cells to a cathartic expulsion of unsuitable sequences. Another source of variation is provided by alternative RNA splicing. In this mechanism, different pieces of RNA are put together in different combinations. This allows the DNA of a single gene to direct the synthesis of many different proteins at different times or in different tissues, as for the brain neurexins and troponins mentioned above.

Yet another way to change the amino acid sequence of a protein is by RNA editing. In this curious process, enzymes modify individual nucleotide bases in RNA, leading to specific replacements of amino acids. One protein of deep-sea squid can be changed at thirteen distinct locations. This protein acts as a channel in the cell membrane of the giant nerve fiber that mediates the escape reflex of the animal. The channel allows potassium ions (positively charged atoms) to pass into the cell, and editing produces subtle changes in its properties. Some amino acid changes alter the clustering of the channel protein in the membrane and influence the number of potassium ions it can carry. Others tune the channel to such features of the marine environment as temperature, pressure, and salinity. Since each site can be changed independently of the others, the number of different channel proteins that are theoretically possible is very large: 8,192. Furthermore, since each channel is composed of four subunits (and these appear to be combined in a random fashion), there should in principle be an astronomical number of distinct channels.

As if this were not enough, proteins are further diversified after synthesis—at the end of the assembly line, so to speak. Cells are full of enzymes that attach small chemical groups such as methyl groups, lipids, sugars, and phosphates to already-made proteins. Some target molecules are modified at several sites, so the potential number of distinct forms increases enormously. Recall the example of glycogen synthase, which can carry up to ten phosphates, described in Chapter 4. Recall the cluster of bacterial receptors with their almost infinite number of distinct methylation states in Chapter 5. Chemical modifications alter the shape and function of a protein just as surely as the binding of an allosteric regulator. So if they lead to an improved survival of the organism, they will be selected and passed on to succeeding generations.

Another dramatic example is provided by histones. These positively charged proteins associate closely with negatively charged DNA to create the compact structure of chromosomes. The basic arrangement of histones is highly repetitive, rather like a series of balls on a string. But it is highly individual in detailed chemistry. The end of each histone molecule extends like a flexible hair to produce a "fur" or "brush" on the chromosome surface. And these histone tails are highly variable, being targeted by multiple enzymes. Something like seventy kinds of histone modification are found in human cells, including phosphate, methyl, and acetyl groups. Since any particular histone molecule can receive more than one modification, the result is a wild proliferation of forms. Histone modifications act like flags on the exposed surface of the chromosome that mark features of the underlying DNA sequence. They may also reflect large-scale organization of the chromosome, such as regions that attach to the membrane enclosing the nucleus. These modifications are sometimes referred to as the "histone code," since they exist in parallel with the DNA sequence itself and supplement the information it carries.

We are led to a striking conclusion. Every small set of interacting protein molecules in a living cell could, in principle, adopt an essentially infinite number of chemical states. But what does this mean? Is it conceivable that every one of these myriad forms performs a different function in the cell, unrepeatable and unique? The answer is no. Many variations in protein structure are likely to result from chance encounters between enzymes and their substrates, a consequence of thermal energy. Others will be caused by trivial fluctuations in the cell's environment that are of no consequence to its survival. But beneath this hubbub of molecular noise there must also be an even tenor of meaningful variation. Any constellation of protein structures that is useful to the cell or organism could in principle be used. If histone molecules next to each other on DNA have a phosphate group here and a methyl group there, then this arrangement might signal an important sequence of bases in the DNA. A second protein could therefore evolve to recognize that particular arrangement of phosphate and methyl groups. The new protein might act as an anchor to attract enzymes that interact with the chromosome in some way.

The extent of meaningful variation in cell structure is completely unknown at present. How many of the variant forms of a given protein result from natural selection? Which modifications to a protein's structure have evolved so that the organism operates optimally under many environmental conditions? We do not know the answers to these questions. But the possibilities are endless.

Thus a basic knowledge of and response to the environment is an integral part of every living cell's makeup. An unending crazy kaleidoscope of environmental challenges has been present throughout evolution. Organisms have responded by changing their chemistry—any that failed to adjust became extinct. And the richest source of variation was in giant polymeric molecules. From a time-compressed view, the sequences and structures of RNA, DNA, and proteins can be thought of as continually morphing in response to the fluctuation of the world around them.

These changes are cumulative: each modification adds to those that have gone before. It is as though as each organism evolves, it builds an image of the world—a description expressed not in words or in pixels but in the language of chemistry. Every cell in your body carries with it an abstraction of some aspect of the world fixed in constellations of atoms. Because of their particular chemistry, enzymes, singly or in groups, respond optimally to combinations of signals from the outside. Like the hidden units of a neural network, they perform acts of classification and abstraction. They are most active—as enzymes, or motors, or signaling molecules—when presented with environmental stimuli of special significance.

The images captured by the cell include everything from recent events to the distant past, rather like a picture taken on a family vacation. In the foreground are the fluxes of ions and small molecules that capture the moment: images of ice cream cones, books, and sunglasses. In the middle distance, protein molecules display evidence of the recent past encountered by the cell in their states of chemical modification and conformational shape: these are the friends, animals, and cars that set one holiday apart from all others. In the background of the composition we have the genetically specified chemical terrain shaped over millennia by evolution—sequences in DNA and structures of proteins that have remained virtually unchanged and are as familiar as shapes, colors, and textures of roads, buildings, trees, and hills.

A living cell contains an image of the world because it is born of the world. From an early protocell three billion years ago to the nerve cells firing here and now in your cerebral cortex, cells have been shaping and refining their chemistry to accommodate environmental pressures. Just as, in the history of mankind, waves of invading armies left their genes to succeeding generations, so the episodes of natural selection have written their stories in the makeup of a cell. Past encounters with physical trauma, food sources, parasites, predators, sexual pressures, are all recorded in the molecular fabric of living cells because, simply, they have had to respond in order to survive. Goethe captured this notion in poetic form when he wrote:

> Were the eye not of the sun,
> How could we behold the light?

It's true: the very essence of evolution. Our eyes are not the product of intelligent design. There was no supernatural draftsman who, taking careful note of the range of frequencies, intensities, durations of electromagnetic radiation, selected this and that molecule, this and that structure. No: our eyes emerged in incremental stages, generation by generation. And at every step of the way their molecules and structures were morphed. First, they had to survive the sun's lethal irradiation. Then, to compete with predators and preys, they had to learn how to use its light in ever more clever ways.

Except that eyes and other sense organs are not single cells. They are built from billions of cells of many types that work together to detect and process external stimuli. In some way the primitive reactivity to the environment intrinsic to individual cells is coupled so that the global responses are extended and amplified. Somehow, it seems that free-living cells, leading an autonomous existence, coalesced into vast communities. Systems of proteins working within individual cells became modified to perform special functions. They acquired the ability to cooperate, producing creatures that far outperform any individual cell. How might this have happened?

NINE

Cells Together

Robert Boyle, seventeenth-century scientist, was settling in for the night in his house in London when one of his servants informed him of a startling finding in the pantry. Apparently, a neck of veal purchased from a country butcher the week before had developed luminous patches of varying brightness and size.

"They made a rather splendid show," he wrote in a paper presented to the Royal Society on March 15, 1672. "In this one piece of meat I reckoned distinctly above twenty several places that did all of them shine, though not all of them alike, some of them doing it but faintly." And later, "The parts that shone most, which 'twas not so easie to determine in the dark, were some gristly or soft parts of the bones, where the Butcher's Cleaver had passed; but these were not the only parts that were luminous; for by drawing to and fro the Medulla spinalis, we found, that a part of that also did not shine ill: And I perceived one place in a Tendon to afford some light."

The spots of light were bright enough to allow Boyle to read several large letters printed on a recently published *Transactions of the Royal Society*. Yet, surprisingly, he found that they created no sense of warmth in the meat. Having to hand the famous vacuum pump or "Pneumatical Engine" he recently had used to study the properties of gases, Boyle applied it to the patches, discovering that the glowing was much diminished in the absence of air. Further inquiry was interrupted by the need to attend a sick niece. "The horrid symptoms . . . being through Gods goodness considerably abated," it was several days before Boyle's thoughts could regain the serenity necessary for philosophical enquiry. Unfortunately, by now the shining veal had been disposed of, but on investigation of the same larder light was seen shining from a pullet hanging in a corner. This meat appeared to Boyle fresh and without obvious signs of putrefaction. He ate it the following day and found it rather good.

The light in Boyle's larder was probably produced by *Photorhabdus,* a species of a bioluminescent bacteria thought to be responsible for curious "glowing wounds" observed during World War I and the American Civil War. The cells of this bacterium create light by converting a small molecule to an unstable form that, in the process of rapid decay, generates light photons. The process, catalyzed by the enzyme luciferase, is extremely efficient. Almost 98 percent of the available chemical energy is converted into light, hence Robert Boyle could detect no heat in his infected meat. Exactly why this organism evolved the ability to make light is still not certain, since it has no eyes or other photosensitive organ.

The relevant aspect of Boyle's lights for us is that the bacteria shine only when they congregate. In other words, luminescence relies on individual bacteria signaling their presence, over a short distance, to each other. Evidence for this curious mechanism came first in studies of another light-producing organism: the marine bacterium *Vibrio fischeri.* In 1970 Kenneth Nealson and John Hastings of Harvard University observed that vibrio cultures do not glow at a constant intensity. In fact, they emit no light whatever until the population reaches a sufficiently high density. Nealson and Hastings already knew about luciferase, which had been detected in extracts of clams as long ago as 1885. They postulated that there must be a molecular messenger traveling between cells. Once the messenger enters a target cell, they suggested, it would induce expression of the genes coding for luciferase and other proteins involved in light production. Their theory met with skepticism at first but has since been confirmed and expanded.

The messenger in this case is a small molecule that does indeed pass from one vibrio cell to another through the surrounding medium. Taken up by the receiving cell, it then stimulates light production, but only if there is enough of it. Light erupts only when the cells are so crowded that messenger levels in the medium (the smell of the crowd, as it were) passes a threshold. At that point, the cell starts making luciferase and its light-producing collaborators. It also starts to make more messenger and to release it into the surroundings. This process creates a positive feedback loop leading eventually to a burst of light.

Why does *Vibrio fischeri* produce light and why only when in dense populations? In this case, unlike Photorhabdus, there are clues. Vibrio bacteria provide the gleam emitted by the light organs of various species of fish and squid. When the bacterial cells live freely in the ocean, they are extremely dilute, and any messenger they release is quickly dispersed; the microbes give off no light. But if they manage to find a safe, nutrition-rich niche in the body of a fish, they multiply rapidly until they are packed like sardines. In fact, specific marine animals develop pockets on their external surface evolved expressly to invite colonization by bioluminescent bacteria. As the bacteria accumulate inside these pockets, the secreted messenger accumulates and triggers maximal light production.

Different marine animals use this light in various ways. A species of squid, as an example, employs its light organs as an antipredation camouflage, to avoid casting a shadow on bright clear nights when the light from the moon and stars penetrates the seawater. A species of fish uses the light produced by *Vibrio fisheri* to attract a mate: two luminescent regions on its face are attractive to fish of the opposite sex. Yet other fish species use their light organs to attract prey or to ward off predators. But whatever the benefit to the host, the advantage to the bacteria is always the same: a safe and nutritionally rich ecological harbor.

By turning on light production only at high population densities, bacteria display a rudimentary social awareness. They are not completely insular; they "talk" to each other. Moreover, the communication channel they use—the secretion and uptake of small molecules—is unexpectedly busy. Cell biologists are only now beginning to appreciate the richness of their language, sometimes referred to as quorum sensing. Not only does it give bacteria a sense of how many other cells are present nearby, it can also distinguish siblings from foreigners, since different kinds of bacteria secrete different substances. Bacteria can also use released chemicals to probe their environment. For example, *Enterococcus faecalis,* one of the most troubling sources of infection in hospitals, deploys a secreted molecule like chemical radar. This substance becomes chemically changed if it encounters a target cell, so if the bacterium finds the molecule "coming back" in modified state like a chemical echo, it knows that food must be somewhere nearby.

Signals sent from one bacterium to another can lead to collective behavior of a kind one normally attributes to higher organisms. There are bacteria that, under stress conditions such as deprivation of food or infection with viruses, commit suicide. The evolutionary logic of this seemingly illogical behavior is that the death of the few can enhance the survival of the many. Its mechanism is once again the secretion of a special substance, in this case a small peptide derived by the breakdown of a common enzyme. Secreted molecules also play a crucial part in the formation of biofilms, through the secretion of long sticky polymers made of polysaccharides and fibrous proteins. Cells in biofilms develop special properties compared with their free-living cousins and often show different specializations in different regions. The resistant plaque that forms on our teeth is one form of biofilm. Others are found in hospitals associated with catheters and prosthetic devices, where they create an important source of infections.

The above examples demonstrate that even though bacteria are individual cells, they are not "islands entire of themselves." They communicate through shared chemicals and thereby expand their capabilities. Consequently, the networks of interconnected biochemical pathways I described in previous chapters do not end at the plasma membrane. They can be connected by specific molecules, sent from one bacterium and taken up by another. The computational possibilities of each bacterial cell are thereby extended to include those of the entire bacterial population.

The relevance to higher plants and animals is made plain when bacteria associate into large multicellular bodies. *Myxobacteria* are rod-shaped bacteria that live in the soil and feed on insoluble macromolecules. They break these insoluble compounds down by secreting enzymes, so it is advantageous for the bacteria to stay together. They multiply in loose swarms, thereby pooling their secreted digestive enzymes (the wolf-pack effect). Within a swarm, individual myxobacteria cells move independently with an unusual gliding motion that does not depend on flagella or other visible surface structures. Yet the swarm containing thousands of cells remains a coherent unit and moves slowly in a persistent direction.

One method of communication between the different cells is through contact guidance. As it moves, each bacterium deposits a trail of slime made of polysaccharides that is followed by other myxobacteria. A second mechanism depends on surface contact between cells—certain molecules on the surface of one cell meet a matching receptor on a neighbor and trigger the latter to move. This "touch and go" system helps keep the cells moving together in a swarm.

Even more impressive cooperation is seen when myxobacteria are starved. Under such conditions, the cells in a swarm cluster tightly, piling on top of one another to build large "fruiting bodies." These develop at the end of a slender stalk, raised above the level of the surrounding colony, and develop into tiny buttons, bright yellow, red, or brown. Within the fruiting bodies, bacterial cells change into spores cemented together by a tough matrix of secreted material. In this state, the spores can survive for long periods, even in extremely hostile environments. Only when conditions are favorable for growth do they germinate to produce normal bacteria. Since the fruiting body is large, and contains a high concentration of spores, a dense swarm is generated rather than a few isolated cells that cannot feed efficiently on their own.

The swarming of individual cells is best understood in another organism with a similar life cycle. The slime mold *Dictyostelium discoideum* is a eukaryotic organism, with larger cells and more genes and molecules than bacteria. Dictyostelium spends part of its life cycle as independent amoebae in the soil, feeding on bacteria and yeast cells and dividing every few hours. When its food supply is exhausted, the amoebae stop dividing and gather in a number of central collecting points, attracted by a chemotactic signal secreted by the amoebae themselves—a sort of quorum sensing. Within each aggregation center the cells adhere to one another by means of special cell surface molecules to form a tiny (1–2 mm) vertical wormlike structure. This falls over on its side and crawls about like a glistening slug, leaving a trail of slime behind it.

The multicellular slug has capacities beyond those of free-living amoebae. It is extremely sensitive to light and heat and will migrate to a light source as feeble as a luminous watch. In its natural environment, a tendency to move toward heat and light directs the slug up from the soil to the forest floor. As the slug migrates, its cells begin to differentiate— undergo progressive changes to a more specialized and usually easily recognized cell type. This process culminates in production of a fruiting body some thirty hours after the beginning of aggregation. Cells in the front of the slug become the stalk region; those behind differentiate into spores; those at the very rear form the footplate. Both stalk cells and the spore cells become encapsulated by walls of cellulose; in the end, all but the spore cells die. Only later, when conditions are favorable, will the spores germinate to produce the single-cell amoebae that start the cycle again.

The concept that plants and animals are built of discrete units—cells— is one of the foundations of modern biology. Formulated in its modern version in the late nineteenth century, this principle of biology has been established beyond doubt by subsequent developments in microscopy, genetics, and tissue culture. And yet it can be taken too far. The cells of a plant are physically connected through their cytoplasms, and cells of both plants and animals communicate freely and abundantly with their neighbors. They send and receive an unceasing barrage of messages and signals that regulate their activities and determine their fate. For many purposes they might as well have no defining membrane.

We can see the part played by cell communication in the small organism *Hydra*, named after the many-headed monster of Greek mythology. This freshwater relative of sea anemones and corals lives in ponds, lakes, and streams, attached to rocks or water plants by a sticky secretion from its base. Looking to the naked eye like a piece of frayed cotton about a centimeter in length, hydra has a threadlike body crowned at one end by a set of six or so tentacles, depending on the species. The tentacles trail, almost motionless, in the water, but their passive appearance is deceptive. Any small crustacean or worm that brushes past is instantly riddled with a shower of poisonous numbing threads. The tentacles then wrap around the paralyzed prey and draw it into a central body cavity called the coelenteron—the hydra's mouth.

It has been known since the work of Abraham Trembley of Geneva in the eighteenth century that hydra cells send signals from one to another along the length of the body and tentacles. Engaged as a tutor-in-residence in The Hague, Trembley occupied his spare time performing experiments on what he termed the "Polyps with arms shaped liked horns." His delicate operations on the tiny animals, described in his *Mémoires* (1744), included the first ever grafts made on animal tissues. In the course of this work Trembley found that separated pieces of hydra could regenerate to form a complete animal. Evidently, as we now realize, cells in the remaining stump can change their specification and position as necessary. Moreover, messages must pass between different parts of the organism so that cells know what they should do at different locations. Somewhere in that small organism there is a body plan, with tentacles and a mouth at one end and an attachment foot at the other.

This plan, moreover, and the signals responsible for its execution are independent of the nervous system. A hydra treated so that it has no nerve cells becomes paralyzed and cannot catch prey. But the animal can be kept alive if food is stuffed into its mouth, and such an animal maintains its body plan and can regenerate any lost parts. It is worth recalling here that single-celled organisms such as stentor also develop and maintain a well-defined anatomy in the absence of a nervous system.

How do hydra cells communicate with one another? This question would require many books of this size to answer. But in broad conceptual terms the requirements are easy to state. In some manner, the signaling pathways of one cell must influence those of another: two networks of biochemical reactions have to become connected. The most obvious way for this linkage to occur is for physical connections to be created between the two cytoplasms. Indeed, some animal tissues such as epithelia do have special junctions that connect cells in this way. But even if adjoining cells remain separated by membranes, protein molecules from the two cells can still form a sort of bridge. Once again, special junctions of this kind are well known.

Communication over longer distances can be also achieved by the secretion of small molecules, as in the quorum sensing of bacteria. In hydra, molecules at one site can diffuse freely in the gaps between cells from one end of the body to the other. Communication here depends on the concept of molecular recognition. Molecules secreted by one cell are detected by binding to receptors on a second cell. Signals are carried across the epithelium from the outside to the inside of the animal by cascades of enzyme-catalyzed reactions. As we have seen, these engage protein machines that act as switches and perform logical functions.

The vastly more complicated higher animals evolved from simpler ancestors resembling amoebae. They arose by the same basic principles of cell cooperation seen in hydra, applied in sophisticated fashion many billions of times. Epithelial sheets of cells cover external and internal surfaces in the body, creating sheltered compartments and controlled internal environments. Within body compartments, specialized cells maintain different functions according to their place. Each tissue of the body has different sets of enzymes and contributes in its own way to the survival of the organism as a whole. Physical and chemical mechanisms then ensure that the different parts work together.

Consider the chemical reactions that result in growth, division, energy production, excretion of waste, and so on. These metabolic processes vary almost as much as those that lead to recognizable products such as hair, teeth, and antibodies. It is true that virtually all cells break down and manufacture glucose, produce energy by combination with oxygen (in mitochondria), make and break down fats and amino acids. But the processes in different tissues are controlled in different ways, fine-tuned in response to the needs of the organism. Nerve cells, probably the most fastidious cells in the body, maintain almost no food reserves. They rely almost entirely on a supply of glucose and fatty acids from the bloodstream. Liver cells, by contrast, make large food reserves—one of their distinguishing features. They supply glucose both to the brain and to contracting muscle cells, recycling the lactic acid produced by muscle cells back to glucose.

Because nerve cells are so dependent on external glucose, it is essential that the supply in the blood is kept constant. The brain needs its glucose day and night, whether we are sleeping, eating, or working. The problem is that during vigorous exercise the glucose consumption of our muscles can increase a thousandfold over their resting rate. When this happens, the liver kicks into action, like an emergency generator, to produce fuel for muscles and the rest of the body. These changes are instigated and coordinated by hormones such as glucagon and adrenaline. Released into the bloodstream, these hormones stimulate the liver's glucose production and mobilize fat from the fat stores. Without them the blood would be rapidly depleted of glucose, and our brain would switch off like a light. The opposite situation occurs after a good meal when we sprawl on a sofa. Glucose from the gut now floods our bloodstream and must be absorbed and stored by the liver and muscle. Another hormone, insulin, now comes into action. Insulin promotes the storage of glucose when blood sugar levels are high after a meal. If this hormone is absent or for any reason fails to work properly, then the serious disorder of diabetes results.

So the body contains not one network of reactions controlling uptake and release of glucose but many. Each cell in every tissue carries its own slightly different and characteristic version. A true representation of the metabolic network would therefore encompass all sorts of cells in different tissues, as well as the composition of the bloodstream. It would include the sources of hormones (the pancreas that produces glucagon and insulin; the adrenal glands that produce adrenaline), as well as an even wider network of signals that control their release.

There's another thing to consider. The most obvious feature of plants and animals, compared with bacteria or amoebae, is their large size and intricate shape. The billions of individual cells of a rosebush or a tiger are not associated in a formless clump but distributed in space in precise positions. The development of shape of flowers, thorns, whiskers, and teeth, known as morphogenesis, progresses in cumulative fashion. As the organism grows from a single fertilized egg, cells pile upon cells like bricks on bricks, building a characteristic architecture. Within this growing structure, individual cells acquire distinctive and usually unique relationships to their neighbors.

Think of a single cell in the skin of your right thumb. This has close contact with similar cells on either side and perhaps also above and below. Any influence that emanates from this one cell will have its greatest effect on just this local cohort of neighbors. If our cell becomes infected with a wart virus, for example, the virus might spread locally to cause local proliferation and accumulation of cells. But a seemingly identical cell in the skin of your left thumb will not be affected. The same will be true of any metabolic influence or regulatory signal that passes from cell to cell through specialized junctions. These also will be locally restricted. So if we attempted to draw up a list or chart to show causal connections between molecular events in our bodies, it would have to somehow embody the anatomy of the organism.

And superimposed on all this we would have to represent factors that act at a distance. These would include hormones such as glucagon and insulin and electrical signals from the brain, carried over long distances by nerve cells. For completeness we would even have to include physical transmission, such as mechanical forces transmitted from one end of a contracting muscle, or between parts of the bony skeleton. The proliferation of links, or connections, in such a network representation would be so exuberant that it would be humanly inaccessible. But this thought exercise does explain why there is such a vast number of proteins and other large molecules—the molecular prodigality referred to earlier. Each cell in the body evidently has to respond to a unique set of biochemical and physiological inputs, tailored to its specific location.

The human eye is often regarded as perfection in design, a beautifully effective optical device made of biological materials. It is also a stunning showcase of cell types, illustrating the variety that can arise from a single fertilized egg. Epithelial cells are there in thin sheets as in hydra, covering the free surfaces of the eye itself and its internal compartments, including lens, iris, and retina. Endothelial cells, similar to epithelia but specialized to regulate the flow of fluids and ions between compartments, line the inner surfaces of the lens and the blood vessels of the eye. The blood vessels ramify within the iris and over the inner surface of the retina, carrying to the eye a mixed population of blood cells responsible for oxygen transport, immunity, and tissue repair. But there are also parts of the eye, such as the cornea and the lens, that lack a blood supply, perhaps to enhance transparency. The cornea features a gel-like matrix of collagen created and maintained by spindly skin cells. The lens, an amazing structure, is built from extravagantly elongated epithelial cells, each a few microns wide and almost a centimeter in length, packed in layers like an onion. The cytoplasm of these cells is rich in globular proteins. They are essentially devoid of the organelles that, in other cells, scatter light and prevent them from being transparent.

Muscle cells are part of the eye. Six conventional muscles, inserted into the eye's outer orbit, enable us to direct our gaze. An unusual circular sheet of muscle surrounds the iris. Reflex movements of the iris open and close the pupil in response to changes in light level and also reflect mental activity such as concentration and pleasure. The iris itself has a rich blood supply and contains cells rich in pigment granules. The color of the iris, whether gray, green, brown, or blue, arises from light-scattering effects of these granules, their size and arrangement. Nerve cells play a crucial role in the eye, not only regulating contractions of its muscles but also as a major component of its retina. Named from the Latin *rete*, net, because of its conspicuous filigree of blood vessels, the retina is more properly regarded as an outpocketing of the brain, a terminal expansion of the optic nerve. This thin sheet of interconnected nerve cells converts light into electrical impulses and carries them to the visual-processing areas of the brain.

Bizarrely, light is actually detected at the very back of the retina. Light has to pass through the web of blood vessels and the fine mesh of nerve fibers, fortunately both transparent, before it reaches a layer of light-sensitive cells. These specialized cells, called rods and cones, possess stalklike extremities full of a special protein. This protein, like many others (hemoglobin is another example), carries a small molecule as a permanent accessory. In this case, the molecular aide-de-camp is a lipidlike compound retinal, related to vitamin A. When it is exposed to light, retinal reacts by twisting into a different molecular shape, and this provokes a larger change in the protein. Biochemical signals caused by light then spread through the rod or cone cell, changing the rate of pumping of ions across the membrane. An electrical signal is generated that passes from this cell to the surrounding nerve net and hence to the brain, and the sensation of light results.

In yet another resonance with our distant origins, the same small molecule is used across the evolutionary canvas. Much to our astonishment, we find that humans, birds, flies, worms, algae, protozoa, and bacteria all employ the same pigment molecule. In every case retinal, cradled in the arms of a membrane protein, is the primary detector of light.

All the different cells of the human eye arise by repeated divisions from a single precursor cell, the fertilized egg. Indeed, from this standpoint, not only the eye but the entire organism is a clone. But it is a clone with a difference: as a plant or animal grows in size, its individual cells change their appearance and acquire novel and highly distinctive properties. Even in the relatively simple hydra, cells diverge into nerve cells, muscle cells, or sting cells. In the human eye there are cells specialized to form a lens, an iris, blood vessels, muscle, nerve cells, rods, and cones. The differentiated features that a cell acquires during embryonic development usually persist into adulthood, even after the initial signaling influences have disappeared. A muscle cell removed from the body and placed in a bath of nutrient medium will retain its properties, including the ability to contract, for long periods of time. Once again, we have to say that a cell has a memory. Once again, we have evidence of complicated logical processes taking place inside cells. But there is one difference: in this case the computations operate at the level of genes.

TEN

Genetic Circuits

In the 1950s a group of biologists led by Jacques Monod and François Jacob working at the Institut Pasteur in Paris found that *E. coli* bacteria could use lactose, a sugar found in milk, as a source of food in place of the more usual glucose. In fact, within minutes of being transferred to a medium containing lactose as the only source of carbon atoms, the bacteria developed the enzymes and other molecules needed for its digestion. This was true even if the bacteria had never previously met lactose; they seemed to have an innate knowledge of this sugar, like an embedded memory.

The most likely explanation for this phenomenon is that at some time in the distant past an ancestor of present-day *E. coli* encountered lactose for the first time. This antique bug went through the slow and painful process of evolving proteins to digest lactose: slow because it needed lucky mutations to produce suitable proteins, painful if we think of the unsuccessful siblings who were outcompeted. Eventually, a family of bacteria acquired the necessary biochemical equipment to deal with lactose. This included a membrane protein for transporting lactose across the membrane into the cell, and an enzyme that breaks down lactose to release glucose. Instructions on how to make these proteins were then stored in DNA and passed on to successive generations.

Lactose is, so to speak, a movable feast, available at some times but not at others. A cell does not need to make the lactose transporter and the lactose enzyme all the time. Indeed, doing so is undesirable, since every protein made by a bacterium is an additional burden, an extra cost. Any cell that can switch off genes that are not being used will gain a competitive edge. This is why, presumably, another protein appeared during evolution able to switch the other lactose genes on or off. The lactose repressor is made in very small quantities—just enough to stop the transporter and the enzyme from being expressed. By investing in a few molecules of repressor, the cell saves the expense of making thousands of transporter and enzyme molecules.

Unless and until it encounters lactose. Then the spring is sprung and the switch is thrown. Repressor protein binds to lactose (strictly speaking, a derivative of lactose), the repression is lifted, and the bug happily devours its newfound food. Through this mechanism, the genetic information for the transporter and the enzyme can be passed from generation to generation. All it requires is the relatively cost-free process of copying strings of base pairs. It is an inherited memory that is utilized to make costly proteins only when conditions are right.

How, precisely, are genes turned on and off? To read the information in DNA, the enzyme RNA polymerase has to copy limited regions containing each gene—in the case of lactose, just three proteins. The new RNA molecule then directs the manufacture of new protein molecules, with the help of ribosomes and the apparatus of protein synthesis. This stream of biochemical events, like a factory production line, can be regulated virtually anywhere along its length. But the most effective point of control is at its source, where RNA polymerase starts to copy the gene.

We have to imagine the bacterial polymerase, a large protein molecule, sliding up and down the length of the DNA until it encounters a promoter. This is a sequence of base pairs upstream from the gene itself that acts as a signpost. Recognizing this sequence through the kind of molecular recognition discussed in Chapter 3, the polymerase binds strongly and switches to an active state. It then moves along the DNA, catalyzing the formation of an RNA copy.

So the short stretch of DNA in the promoter region is crucial for control. Mutational changes to the base pairs in this region have a direct impact on the strength of binding of the RNA polymerase and hence on how efficiently it makes RNA. Moreover, if another protein in the cell independently evolves a site that recognizes this same sequence—sits on precisely the same region of DNA—then it can affect the binding of the RNA polymerase. This is how the repressor controls genes necessary to digest lactose.

This simple genetic switch, known for more than fifty years, is an object lesson. It shows us in principle how cells select which parts of their DNA to use. Many other regulatory circuits composed mainly of proteins (but also including in some instances RNA, small molecules, and DNA itself) control whether and to what extent a particular gene is turned on. Once again it's easy to see analogies with electronic or engineering systems. And given the ease with which one can perform genetic manipulations in bacteria, it is tempting to imagine that one might go into a cell with the equivalent of a soldering iron and change the wiring.

Now the repressor is a protein like any other and therefore made according to instructions encoded in DNA. It also has a promoter site that can be modulated by other repressors or activators. These are yet other proteins. So there is a possibility of an extended regression . . . an endless cascade of processes, with one gene influencing another gene that influences a third gene and so on. Even the simplest genetic circuit designed on this principle can create lively dynamics.

Imagine a push-me-pull-you circuit where protein X represses gene Y and at the time protein Y represses gene X. What do you expect will happen? Yes, a struggle for supremacy. If X gains the upper hand, even briefly, then it will shut down Y. As the amount of Y falls, even more X will be made, and so on. The system should slide irresistibly to an all-X state or—since exactly the same argument holds if protein Y gets its foot in the door first—an all-Y state.

And indeed it does, as shown by genetic circuits designed and made to order. In 2000 a team from Boston University led by the engineer James Collins put together one of the first designed genetic circuits—an arrangement of two genes similar to the XY case above—in the bacterium *E. coli*. As Collins and his colleagues recounted in the journal *Nature,* two repressible promoters were arranged in a mutually inhibitory network and, as expected, behaved like a toggle switch. The bacteria synthesized just one or the other of the proteins, in stable fashion. Both promoters were protein switches, and each became active when it bound a specific small molecule (just as lactose induces the lactose genes). Exposing the bacteria to one of these small molecules was enough to send the system irreversibly into one of the two states: it made just one of the two possible proteins.

With the synchronicity that so often accompanies scientific developments, another designed circuit appeared in the same volume of *Nature*. Stanislaw Leibler, a physicist turned biologist originally from Poland, and his graduate student Michael Elowitz, working at Princeton, reported the creation of an oscillating circuit in bacteria. In this case, they introduced three genes into the bacterium, each controlled by one of the others. Rather like the traditional children's game of rock-paper-scissors, in which paper beats rock, scissors beat paper, and rock beats scissors, the product of each gene in this triad repressed the next in the circle.

The operation of the cycle was detected by an additional genetic construct that made a fluorescent protein whenever one of the three genes was at its lowest concentration. Because three genes were linked rather than two, and because the genes had suitable strengths of binding and protein lifetimes, levels of the three proteins rose and fell continuously. Under the microscope, the bacteria looked like a population of lights turning on and off. But slowly: the average period between flashing being around 150 minutes. Since this is about three times longer than the time the cells took to divide, the state of the oscillator circuit was evidently being transmitted from generation to generation.

These humble bacteria, modified by standard techniques to achieve a predictable result, had an electrifying effect on contemporary biologists. It was a proof of the principle, you see, that one could design genetic circuits to order. A flurry of papers appeared using similar methods. Within a year, an undergraduate course in synthetic biology at MIT presented students with a tool kit of genes and encouraged them to build novel circuits.

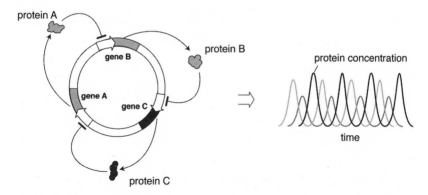

FIGURE 10.1. A genetic oscillator. Circular DNA molecules carrying a set of three genes (shown on left) are introduced into bacteria. Each gene makes a protein that inhibits its next-door gene. Because of time lags between making a protein and its having an effect on the next gene, the system settles into an oscillatory mode. The concentration of each protein then rises and falls in periodic fashion, as shown on the right. Based on Elowitz and Leibler 2000.

Consider the possibilities. If you can make an oscillator, then you should be able to add simple logical elements to control its phase and period: when the lights go on, and for how long. For inputs you might use small molecules such as amino acids or sugars added to the culture. And since this is possible, why not modify the bacteria so that they emit a light signal when they detect a pollutant, say, or a toxin in their environment? Why not include quorum sensing in the circuit? Could one make populations of bacteria blink in synchrony; generate geometrical patterns; propagate waves?

The actual realization of these circuits, it transpired, was far from simple: easy to draw on paper but requiring many weeks of careful, expert work to have any hope of working in reality. The performance of a genetic circuit depends on hard-to-control factors such as how tightly the repressors bind to their promoters, how fast different proteins are made, how long-lived they are, and how stable their RNA molecules. Then there is the question of noise. Living cells vary enormously in size and the number of molecules they contain, and this has a major effect on the performance of genetic circuits. These biochemical realities place severe limits on the numbers of genetic elements that can be usefully combined. To date, by most criteria, these handcrafted biological circuits are pitifully inaccurate and primitive when compared with silicon devices.

But they have one unique advantage: they reproduce themselves. Suppose you have designed a genetic circuit and successfully installed it in a small sample of bacteria. How could you make more? Why, simply add nutrient broth! Within a day or so, each one of your original bacteria will have generated a billion offspring, each carrying a copy of the circuit you designed. This remarkable achievement has nothing to do with your cleverness but simply reflects the intrinsic property of living things to reproduce themselves, in all their parts. The new genes that you added to make your circuit are simply swept up with the general machinery of reproduction.

Gene circuits do not exist in isolation. They are embedded in a living cell and are able to function only because of all of the support they receive. They need a warm, aqueous environment supplied with the right mixture of salts, ATP, and other energy-rich compounds. They need a steady supply of raw materials, together with all the machinery needed to manufacture molecules of RNA, DNA, and proteins. When viewed from this perspective, the term *synthetic biology* seems premature, even boastful. We are a long way from being able to make a complete system from scratch.

There is an analogy here to the writing of computer programs. Lines of software code can produce amazingly intricate patterns on a computer screen. But they do this only because they are supplied with a huge infrastructure of supporting code (libraries, headers, compilers), as well as the physical apparatus of the computer itself.

So how do designer genetic circuits compare to the authentic, homegrown variety? As you might expect, the naturally occurring circuits in living cells are more convoluted and enigmatic: after all, they are the product of evolution, not of intelligent design. Even the way that bacteria control their lactose genes, the archetype of genetic control, is substantially more complicated than my account. In addition to the repressor responsible for shutting off genes, there is an activator that turns them on. The activator in this case is a protein called CAP that binds to the small molecule cyclic AMP. When the bug is hungry, more cyclic AMP is made. AMP binds to CAP and activates genes the bacterium needs to utilize other sources of nutrients, such as lactose.

Incidentally, the role of cyclic AMP as a hunger signal is another of those curious resonances of cell chemistry. This same small molecule in the human liver also rises in response to starvation and falls when we are well fed. The mechanisms of action in the two situations are entirely different. In the liver, cyclic AMP rises in response to the hormone glucagon, made in the pancreas and delivered via the bloodstream. Cyclic AMP forms part of a cascade of reactions in a liver cell that leads eventually to the degradation of the energy store glycogen and the release of glucose into the cell (Chapter 6). So the biochemical details of how this small molecule operates are entirely different in the two situations. It is hard to imagine a series of evolutionary steps that would connect these two pathways, one in a bacterium and the other in the human liver. But the plain fact is, this small molecule signifies "hunger" in both situations.

Coming back to lactose, we see that the expression of lactose genes in bacteria requires two conditions to be fulfilled. First, the cell should be starved of other nutrients, especially glucose. Second, the sugar lactose must be available in useful amounts. Hunger causes high levels of cyclic AMP and turns the activator on, while high levels of lactose turn the repressor off. Nor, indeed, is this the end of the story. It turns out that the repressor has a choice of where it binds to DNA, since there are several repressor-binding sites. Although one site exerts the greatest effect, the others are required for full repression. Moreover, the precise binding strength of repressors for sites on DNA changes rapidly with the environment. Bacteria grown in the presence of different levels of lactose modulate their repressor binding strength within a few hundred generations. In this way they tune expression levels to the precise quantity of sugar available for digestion—no point in making more enzyme than you need.

Bacteria use many other tricks and dodges to control their genes. Repressors and activators are combined in clever ways: summed, subtracted, divided, or multiplied, depending on their precise placement on the DNA. Other regulatory proteins act by bending the DNA physically, changing its twist so that it becomes accessible to other proteins. These are indeed clever cells, and it is fascinating to unravel the intricacies of their internal wiring. But the interior of a plant or animal cell is an even more astonishing place to visit.

Imagine yourself a protein in a cell, part of a complex attached to the inner face of the membrane. Initially quiescent, you are suddenly galvanized into action by signals spreading through the cytoplasm. The cell is under attack by a virus! As alarm molecules find you, you switch into a new leaner and meaner form. Releasing your holdfast, you diffuse into the surrounding hubbub of the cytoplasm . . .

Within seconds you find your way into the cell nucleus. This is a different world, with enormous snakes of DNA coated with proteins coiling away in all directions. Moving at headlong pace, you slip along the sinuous highways this way and that until you abruptly encounter a region where the smooth highway surface ends. Switching to low gear, you move methodically, searching along the rough unmade bed of atoms . . .

Finding at last a perfectly patterned set of atoms, you wriggle into place. Other proteins, following the same route, slide in beside you. The huddle grows larger, settles into a new architecture. Suddenly a behemoth appears, sliding along the highway. As it collides with your new community, parts fly off and massive internal changes propagate through its structure. The newcomer metamorphoses into a giant machine that digs into the coils of the DNA highway, forcing them apart as though with a lever. Your short-lived community of molecules breaks up, and you are released. As you diffuse back to the membrane, the machine begins its methodical progression along DNA, making RNA.

Fantasy, but something like this is happening now in your body, a billion times every second. If you had to point to a single factor to account for the enormous difference between plants and animals compared with bacteria, it would be the extent and richness of their genetic circuitry. Almost 10 percent of the genes in mammalian DNA—several thousand kinds of protein in a human—are directly involved in gene regulation. Responding to signals from the environment and, especially, from other cells, they create skin, muscle, blood, and bone—that is, they build a body. At a glance, these proteins seem to operate like bacterial repressors and activators. But look closer and you will see that they have a far richer repertoire. Some act as boundary markers to demarcate a region of DNA; others modify the three-dimensional arrangement of the DNA itself, causing it to curve or bend back on itself. There are gene-regulatory proteins that displace histones—the positively charged proteins that pack DNA into chromosomes. Others modify histones at specific locations, marking them as targets for other proteins.

And although I focus on protein computations in this book, RNA molecules also have a place. A network of small RNAs has been uncovered inside eukaryotic cells that participate in the regulation of genes. Some operate in the nucleus, working alongside transcription factors; others control the processing of messenger RNA by alternative splicing (see Chapter 8). An even larger population of very short RNAs operate in the cytoplasm, where they influence, often in an inhibitory manner, the translation of messages into protein.

The control regions of eukaryotic genes are much larger than those in bacteria and may be scattered over thousands of DNA nucleotides. There are many control sites and many proteins and RNA molecules to interact with them, working in a promiscuous and overlapping fashion. One protein might influence a large number of genes and synchronize their performances, like the conductor of an orchestra. Except that in this orchestra the players can play more than one tune and may follow many conductors at the same time. Large protein complexes assemble from regulatory proteins, often at specific sites on the DNA. Their activity depends on which molecules they contain but can also be modified after they are built. Enzymes add and remove phosphate groups, changing protein shapes and activity. Small RNAs come and go. Clusters of proteins grow and influence each other, sometimes indirectly, by bending neighboring DNA.

Although changes in expression are often triggered by transient signals, such as a sudden rise in calcium ions or the release of a hormone, their effects can be long-lasting. Cells in an embryo start off as naïve generalists, jacks-of-all-trades that do many things but none very well. They finish in the adult body (with a few exceptions, such as stem cells) as dedicated specialists. Like concert pianists or Grand Prix racing drivers, they are now professionals who do one job very well. And as they become progressively more specialized, cells make an increasingly restricted set of proteins. Each genetic switch builds its effect onto those of a previous one, consolidating its action.

A strange way to make a circuit, you might say. Yes, it is digital in the sense of positive and negative switches. But to say this is a superficial caricature. Genes can be activated to different degrees, with a higher or lower probability. They respond to a plethora of controls from many different sources and employ many different mechanisms. Although some proteins serve as simple switches, many others work in groups, rather like a committee that decides when and where to express a particular gene, and to what extent. It is a long way from a silicon chip or any circuit a human would design.

There are about twenty-five thousand genes in a human cell—comparable to the number of products on the shelves of a supermarket—each gene subject to the subtleties of regulatory mechanisms just described. An adult human body has perhaps fifty trillion cells. The computational possibilities are immense. Is there any way to predict global properties of such a large and heterogeneous population of computational elements? How far can you go without knowing all, or even most, of the individual connections?

In the fall of 1967 a young scientist named Stuart Kauffman made one of the earliest stabs at this problem. He was spending three months at the Research Laboratory of Electronics at MIT, supervised by Warren McCulloch, who had earlier played a leading role in the development of neural networks (see Chapter 6). McCulloch had continued to explore the interface between artificial networks and biology and the young

Kauffman, on his arrival at MIT, took up the issue of gene circuits. At this time the broad features of gene regulation in bacteria were established, but next to nothing was known about how eukaryotic genes were controlled. Undeterred, Kauffman developed the notion of idealized "genes" as binary devices. These existed in one of two states, either on or off, and were linked randomly in large networks.

In a paper published in the *Journal of Theoretical Biology,* Kauffman showed that even a random network of genes, each controlled by two or three others, behaves with a surprising degree of order and stability. Closed loops of activation become established as sets of genes became entrained with each other. With the breathtaking audacity that was to characterize his career, Kauffman drew parallels between these closed loops and the several hundred different types of cells found in higher organisms. His idea was that the distinctive features of a liver cell, a heart cell, or a nerve cell arose because of unique sets of genes linked in cycles of mutual activation. Kauffman also found that transitions from one stable state to another could be achieved by adding noise to the system, likening the process of cell differentiation in an embryo to a random walk among genetic modes.

Over the ensuing decades, Kauffman pursued his vision with missionary zeal, from cycles of interacting genes to the origins of life, social dynamics, and the stock market. Moving to the far-sighted, far-out Santa Fe Institute, set in a brown stucco house in the high desert of New Mexico, he became a guru in the controversial field of "complexity research." In public talks, journal publications, and popular books, he expounded on the abilities of interacting networks of all kinds to self-assemble, to show spontaneous order. In the preface to his 1995 book *At Home in the Universe,* he proposes to show that:

> laws of complexity . . . govern how life arose naturally from a soup of molecules, evolving into the biosphere we see today. Whether we are talking about molecules cooperating to form cells or organisms cooperating to form ecosystems or buyers and sellers cooperating to form markets and economies, we will find grounds to believe that Darwinism is not enough, that

natural selection cannot be the sole source of the order we see in the world. In crafting the living world, selection has always acted on systems that exhibit spontaneous order. If I am right, this underlying order, further honed by selection, augurs a new place for us—expected, rather than vastly improbable, at home in the universe in a newly understood way.

The difficulty with universal prescriptions of this kind—the reason that they remain stranded in the reedbanks of mainstream science—is that they are incorrigibly abstract. It is a seductive notion that a few key ideas might equip us to roll the natural world into a ball. Kauffman (who has since moved to the colder climate of the University of Calgary) and his successors at the Santa Fe Institute talk with high excitement about adaptive systems, conveying a sense of euphoria and omnipotence. But in truth the practical consequences of their theories are small in number.

Universal theories that treat the components of a biological system— proteins, genes, cells, neurons, or people—as identical ciphers rarely lead to specific predictions or applications. They teach us what happens when large numbers of computational elements are linked according to certain rules. The results may be unexpected. But far from revealing the secrets of life, these patterns show us what living things have in common with the rest of the universe. They exist by default and are immediately abrogated if there is any selective advantage to be gained. The more we learn about living systems, the more we realize how idiosyncratic and discontinuous they are. Biological networks are accretions of special cases. Elegant mathematical formulations that homogenize all of the diverse elements are fated to be vague generalizations.

Forty years on from Kauffman's initial proposals we are at last uncovering the actual mechanisms of cell differentiation, especially in a few accessible organisms such as sea urchins and fruitflies. It is now clear that the control panel of gene expression is indeed built from banks of gene switches. In a living embryo, protein activators and repressors, collectively termed transcription factors, operate in cascades. Small RNA molecules, copied from regions of DNA previously thought to be meaningless, modulate activities in both the nucleus and the cytoplasm. Control elements made at higher levels throw switches lower down. Patterns develop over time. Specific genes become expressed in subpopulations of cells as the embryo develops. They become locked into mutually active pathways by positive feedback mechanisms. In a general sense, therefore, Kauffman was correct when he proposed in the 1960s that genetic modules define stable differentiated cell types such as muscle and nerve.

But the networks of genes in any real organism have a richness and baroque strangeness that would be difficult to invent. Fact is stranger than fiction, in cell biology as in human affairs. Consider the early stages of development of a sea urchin, one of the simpler forms of animal life. Thanks largely to the efforts of Eric Davidson, a biologist of similar vintage as Stuart Kauffman, the serial activation of fifty or so genes in the sea urchin larva has been uncovered. Davidson's passion is for experiment rather than theory. Over several decades and with a large team of collaborators (it takes many more hands to perform experiments than to develop computer simulations), he has pieced together the core genetic circuitry of an especially simple and well-defined developmental system. The result is more complicated and idiosyncratic than anyone could have imagined.

Early development of the microscopic swimming larva of the sea urchin is dominated by a set of genes encoding transcription factors. These are expressed at specific times and places in the embryo and operate on clusters of DNA sequences. Most transcription factors increase the probability of gene expression. But there are also those that turn off specific genes, or insulate large regions of DNA one from the other. A typical regulatory region in DNA may be five hundred base pairs in length and contain sites for four or five transcription factors: a typical gene might have three or so such controlling regions. Transcription factors are frequently promiscuous and interact with multiple sites. The result is a network of interacting genes that grows in size and numbers of connections with the embryo.

Each cell in the embryo has a highly individual state determined by billions of protein switches and, like a human face, bears evidence of both

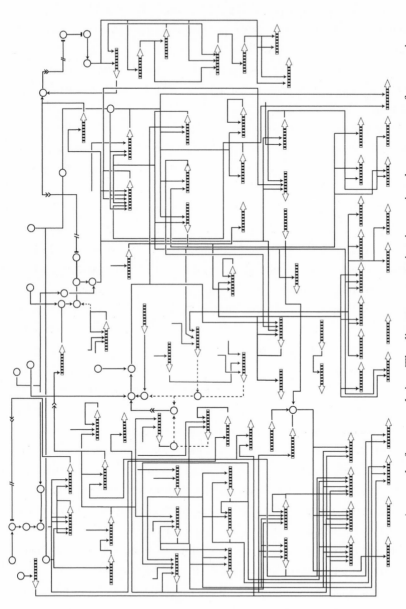

FIGURE 10.2. A network of gene regulation. The diagram summarizes interactions between a set of genes active during a particular stage of development of a sea urchin. Individual genes are shown as banded arrows, their names omitted for clarity. Based on Levine and Davidson 2005.

nature and nurture. *Nature* here refers to the inheritance of differentiated characteristics present in the mother cell. As the embryo develops, different lineages of cells become committed to particular paths of development, increasingly specialized and canalized. Their differentiated character passes to daughter cells on division. It is a form of cell memory that resembles in some respects genetic inheritance and is sometimes referred to as epigenetics. But superimposed on this inherited state are nurtural influences from the environment. Signals diffusing in from the outside make other changes to the constellation of switches. Most of the signals come from neighboring cells in the tissue and themselves contain information on their specialization. Specification and consolidation of these paths will engage further cascades of genetic switches.

In Chapter 4 I described living cells as being crammed full of protein molecules. Acting individually or in small assemblies, they perform reiterated molecular processes that can be regarded, I argued, as a form of computation. Moreover, large numbers of proteins linked into huge interacting networks operate, in effect, like circuits of electrical or electronic devices. Networks of this kind are the basis for the animate wanderings of single cells and their ability to choose what to do next.

In this chapter I have broadened the view to encompass multiple cells—"societies" of cells. Through a variety of strategies—including diffusive hormones, electrical signals, and mechanical interactions—the computational networks of individual cells are linked. During evolution, cells acquired the capacity to work together in social groups; it became advantageous for most cells to become highly specialized. Liver cells, muscle cells, skin cells, and so on abandoned their opportunities for unlimited replication. They began the communal expansion of interlinked abilities that led to the plants and animals we see around us today. But the basis of this diversification of cell chemistry was yet another form of computation—one that operates on DNA. Control mechanisms, again based on protein switches, created extensive but subtle modifications of the core genetic information. The working out of intricate cascades of specification and interaction as an embryo grows leads to highly differentiated cells and organs.

The computational functions of living tissues are integral to every aspect of life—including this growth and differentiation of form. But their most mysterious and alluring activity is conducted inside nerve cells, within the cranial space of our brains. Thanks to this irregularly shaped, gooey lump of organic substances that is 70 percent water, I can walk, talk, remember, estimate, calculate, experience, and feel. Somehow, this apotheosis of wetware has grown from the same fertile compost as other features of living organisms. It is built from protein molecules like any others, individually performing antlike switching tasks oblivious of their role in the higher scheme of things. But the computations performed by this molecular hardware achieve a higher level of sophistication. They underpin the information-carrying events of the nervous system and the cognitive processes of the brain.

Earlier in the book I referred to the dialectic between electronic devices and biology (Chapter 2). I discussed parallels between the actions of individual protein molecules and transistors (Chapter 3); the possibility of mimicking essential features of single cell behavior via simple computer programs and rudimentary mechanical devices (Chapter 5); the ability of proteins to operate like neural networks to recognize patterns and perform abstractions of their inputs (Chapter 6). The interplay was informative and, I hope, interesting. But the replication in machine form of human intelligence and thought processes is an altogether more personal affair. The dream of creating human life is deep in our psyche and has fascinated thinkers for centuries. Its roots lie, perhaps, in our horror of death and desire for immortality. We have made huge strides in recent years in the areas of artificial intelligence and humanoid robots, and this progress challenges our ideas of how the brain works. Our success or failure in creating these alter egos is not only a measure of how well we understand living processes. It also reveals how well we know ourselves.

ELEVEN

Robots

"And this is Lucy," Steve announced, as I entered the workshop, stretching my legs after the long drive through the Somerset countryside. But before I could set down my bags and after the briefest of greetings, Steve Grand and his wife, Ann, waved me to the outbuilding. Situated to the right of their small house, set under one of the trees in their leafy garden, was a single-story outhouse, probably built as a large garage or perhaps a granny annexe. I entered a single spacious room, low-ceilinged and filled with workbenches. Electronic gear was everywhere—transistors, small motors, cogs, wires, soldering irons—piled in meaningful chaos, as though caught by flash photography in flagrante delicto. Computers sat on tables and under benches. Some had their covers removed, with coils of cables disappearing into their entrails, evidently adapted to tasks beyond their designated function. And in central pride of place, facing me as we entered through the door, was a diminutive waif made of struts and wires. Lucy glared balefully at me, her single optical lens focusing noisily as we approached.

These were early days in the Lucy project. At this stage she had only the sketchiest humanoid features. Yet to come were the toweling cover to the face and the cosmetic false eyes that would eventually become familiar to thousands of Internet fans. As we saw her then, she was not much more than a metal frame, about two feet high, with a head and two flexible robotic arms. Her internal anatomy was painfully evident: a mare's nest of silicon chips, relays, and electric motors within her metal skeleton. The mobile head full of highly visible wiring sat on a metallic neck, its features delineated by a crude mask of metal straps.

Steve and I had first met a year previously when we were speakers at the same scientific meeting in 1996. The maverick molecular biologist Roger Brent had organized Beyond the Genome to address the future of biology. The complete DNA sequences of a number of species had just been determined; the human genome was expected within the year. It was time to take stock, to assess how far we had come . . . to speculate on what a future brim-full of quantitative data would bring. Far from being a conventional scientific symposium, this one housed an eclectic collection of scientists from academia, government labs, and industry. In addition to the expected list of geneticists, biochemists, and bioin-formaticists, a woman from the U.S. Air Force talked about image analysis methods that could identify an ammunition dump from fifty

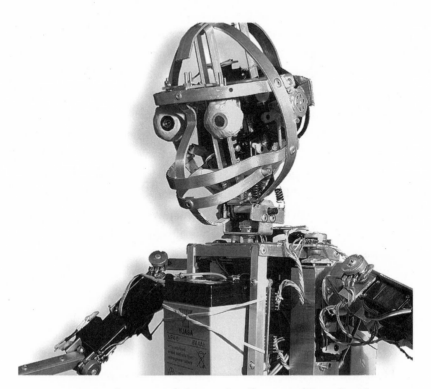

FIGURE 11.1. Lucy, an early incarnation. Courtesy of Steve Grand.

thousand feet. An ecologist regaled us with radical notions of how to counter carbon accumulation in the atmosphere by sinking vegetable waste deep in the ocean.

Arguably the wackiest and farthest out of the speakers was Steve Grand. At this stage in his career, he had the demeanor of a young businessman. A professionally suited technical director of a software company, he was also their senior programmer—creator of a tribe of software creatures called Norns. Purchasers of this software (still available, although Steve is no longer part of the company) acquired on their computer a virtual environment, a house with multiple rooms upstairs and downstairs, set in rolling countryside. Built in what might be called Disney Gothic, the house was fully furnished complete with everyday household items. It had its own small garden with flowerbeds, tool shed, and swings. The house was initially uninhabited, but several eggs were supplied with the initial purchase. Once installed, these developed into Norns—tiny creatures that grew and developed a range of humanlike qualities.

Norns were indisputably Beyond the Genome . . . way beyond! Steve's talk was novel and provocative, but its closest relevance to biological science was in the level of detail—the verisimilitude—he had programmed into the simulations. The diminutive Norns were anatomically complete, with arms, hands, and feet. Their internal chemistry was like a strip cartoon version of a biochemical text, complete with three hundred or so "genes." In his talk, Steve described the health checks he applied to ailing Norns. Graphical displays monitored current levels of glucose, glycogen, and amino acids, as well as proportions of muscle and adipose tissue in the body—making it easy to spot an obese Norn that was eating too much. A handful of hormones, loosely based on human endocrinology, governed the Norns' daily emotions and actions and their nightly sleep patterns. Most detailed of all, appropriately, was the nervous system and brain of the little creatures. This was built from a complicated network of neuronlike elements, able to respond to external stimuli and internal biochemistry. They produced distinctive patterns of behavior that was idiosyncratic and unpredictable. Each Norn was indeed an individual and had characteristics that quickly became recognized, sometimes loved, by its owner.

Writing software can be a highly individualistic and creative activity. In the myriad choices and decisions that went into the 250,000 lines of code for the Norns, Steve stamped his view of the world. An unorthodox loner, largely self-taught, Steve has a clear-eyed inquiring mind that comes to problems afresh, confronting fundamental issues with directness and lack of embarrassment. Some find his distinctive approach naïve, others consider him a genius. Although he is largely ignored by mainstream scientists, a national British newspaper nevertheless voted Steve as one of the "brains behind the 21st century," and his Norns have become iconic figures in the world of software games.

At the core of Steve's philosophy is the conviction—later expounded in his 2000 book *Creation: Life and How to Make It*—that the world is made of events and relationships rather than things. He applies this credo to the entire scale of the natural world, from subatomic particles, existing as waves, through atoms and molecules to networks of interacting macromolecules, cells, minds, and even societies. Living organisms, he declares, are made not of atoms and molecules but rather of cycles of cause and effect. They happen to *use* atoms and molecules, but this is not essential. Stuart Kauffman enunciated a similar view in regard to networks of gene expression. Life, at its root, he said, "lies in the property of catalytic closure among a collection of molecular species. Alone, each molecular species is dead. Jointly, once catalytic closure among them is achieved, the collective system of molecules is alive."

Despite repeated boyish, self-deprecatory admissions of inadequacy scattered throughout his writings, Steve Grand displays the unshakeable personal conviction of a prophet. In *Creation* he refers to his software Norns as "the closest thing to a new form of life on this planet in four billion years." Comparing himself to Victor Frankenstein, he says: "We too are going to use electricity to create life, but our electric current will animate a digital computer." These jaw-dropping claims arise from his view of life as systems of dynamic relationships rather than things. He is able to reproduce the essence of life by reproducing the appropriate interactions on a computer, especially because—as he explains in *Creation*—he embeds simulations at one level within simulations at another. Although he accepts that representations at the lowest level are obvious software fakes, he insists that those at the top of the hierarchy are not. The actions of Ron—his first Norn—constituted an emergent phenomenon that was not explicitly programmed and could not easily have been predicted. Therefore, Steve says, Ron acquires an inalienable claim to reality. He admits that the individual nerve cells in Ron's brain are nothing more than simple integrative circuits with conditional links to their inputs and outputs. But the ensemble of nerve cells, working within the framework of a simulated body with legs and arms and internal biochemistry, becomes "a brain rather than a computer program, and it does in a rather limited sense think."

Defining what you mean by "life" is a slippery business even for a professional biologist. Do you restrict the term, as some do, to complete organisms capable of independent growth and replication? Or do you include parasitic organisms that exist only because they inhabit the bodies of larger organisms? If you say yes to the latter, then what about viruses? In their infective form these are inert, crystalline assemblies of dumb molecules, but in the cytoplasm of a host they metamorphose into greedy self-replicating aliens. And if you will allow viruses, then what about mitochondria, centrosomes, and other organelles capable of directing their own replication? What about plasmids—simple loops of DNA that in suitable environments can replicate and undergo something very like mutation and selection?

Whatever your stance on these matters you will probably feel, as I do, that computer programs—spurts of electrons, weaving patterns on a silicon chip—are in a different category. The fact that there are millions of software Norns in the world and that some owners become emotionally attached to their charges does not make them any more real. Trapped in the pixelated two-dimensional confines of an LCD screen—utterly dependent on the provision of an electronic computer and a human user—these virtual creatures belong firmly in the world of artificial, not real, life.

Since Lucy, by contrast, exists in tangible corporeality, this may explain why Steve abandoned his two-dimensional software offspring for robotics. Turning his back on the commercial world, Steve retired to his house in Somerset. There he struggled to embody his ideas about dynamic loops of brain activity and the contextual requirements for the development of intelligence in an actual robot. Results so far, as outlined in his 2003 book *Bringing Up Lucy,* are undeniably modest. The peak of achievement to date is apparently that Lucy was, once, able to recognize a banana. For the most part, working essentially single-handedly, and beset by severe financial and other practical constraints, this noble enterprise has so far failed to lift off.

And Steve has serious competition. Predictions by sci-fi writers of the early twentieth century that robots would someday become part of our daily life seem now close to fulfillment. A billion-dollar industry exists to provide us with machines able to perform tasks previously the preserve of humans. The scope is wide, and the boundary between possibly intelligent robots and sophisticated automata is often hard to define. Articulated arms in a car factory production line seize incomplete parts of the engine or chassis and assemble them with superhuman delicacy and precision. They are termed industrial robots because they have a superficial similarity to, and historically were inspired by, the human arm. But they are in other respects no different from other factory machines. Conversely, automated bank tellers are not normally thought of as robots even though they perform lightning calculations with absolute accuracy. Sometimes they even speak and display an image on a screen like a real teller. Industrial and commercial devices of both kinds proliferate in our world. They may perhaps once have been informed and inspired by biology, especially human biology. But no one now believes they are in any sense living.

Robotic toys such as the CyberDog are somewhat closer to biology. Designed to interface with a human, they provide as realistic an appearance as seems feasible. Some even can talk and carry out conversations on everyday matters. Creatures such as these seem destined to play an increasing role in our society. In Japan, a conservative, aging population is traditionally suspicious of immigrant labor but appears to be at ease with artificial forms of life. Paro, a fluffy seal toy that responds to pats and strokes with evident pleasure, was developed to help care of the elderly. Paro triggers feelings of emotional attachment and is claimed to greatly benefit the psychological state of mentally frail elderly. Coincidentally, it also monitors daily activities and reports signs of distress to attendant staff. Other enterprises under development include domestic robots that cut the grass, clean the house, monitor household safety, and prepare simple meals. In all of these tasks, the robot acts for long periods independently of human intervention.

It seems widely accepted that humanoid machines will populate our future. Witness the fact that discussions are already under way on the ethical treatment of robots. In 2007 South Korea's Ministry of Commerce, Industry, and Energy established a Robot Ethics Charter designed to prevent humans from abusing robots and vice versa. Similar concerns are being voiced in Japan, the United Kingdom, and the European Union. If official bodies find it necessary to ask whether mechanical devices should be accorded similar rights to those of, say, living pets or farm animals, then they presumably accept the possibility of these synthetic creatures experiencing sensations, emotion, and pain. It seems that they anticipate the development of sentience in these creatures. But how close are we to achieving this?

We need to look elsewhere to answer this question. Mass-produced commercial robots such as Paro and AIBO may be made to look, cosmetically, as realistic as possible. But their manufacturers have little incentive to reproduce actual biological mechanisms beneath the skin. They surely have few illusions about the awareness or sentience of their products, however deluded their customers might be.

It might be that researchers in robot laboratories have a different motivation. Like Grey Walter and Steve Grand they often aspire to create machines that not only *behave* like organisms but also in a sense *are* alive. This, at least, seems to be the underlying ethos of the Humanoid Robotics Group at MIT directed by Rodney Brooks. Over the past twenty years Brooks has built and inspired others to build an entire menagerie of innovative robots. His Web site displays a stable of contemporary robots such as Cog, Kismet, and Coco that, like Lucy, are targeted to such high-level human abilities as visual recognition and speech. Retired robots on the list include six-legged, clambering Genghis, Hannibal, and Boadicea (designed with planetary exploration in mind). Prototypical can-shaped Frankie, Toto, and Allen navigate unfamiliar office environments on wheels. Featured in articles, books, television, and at least one movie, Brooks's robots are among the best-known specimens of their species, at least in the Western world.

The six-legged robotic insect Genghis achieved the iconic status of a display at the National Air and Space Museum in Washington, D.C. Genghis uses infrared sensors to detect the heat given off by nearby humans, whom it then pursues with the single-minded ruthlessness of its namesake. Video clips display it clambering over and around obstacles in its path with an unmistakably beetlelike gait, thanks to the semiautonomous action of its legs, a built-in tiltometer, and two sensory whiskers. Genghis's circuitry—extensively described in Brooks's 2002 book *Robot*—consists of an assemblage of fifty-seven tiny computers, each performing a limited set of tasks in response to simple instructions relayed from other units. One set of microcircuits controls the forward-back motion of legs, another their up-and-down motion. Others process data from the heat sensors, tiltometer, and whiskers.

Development of Genghis followed a bootstrapping approach. Successive levels of performance—standing, walking, clambering, detection of prey—were added one by one, each being built onto a previous already-functional level. This empirical strategy, reminiscent of natural evolution, represented a radical departure from the symbolic-processing approaches to robot design that had been the focus of artificial intelligence since the days of Alan Turing. Genghis navigates and performs its designated task without a central controller. Rather than attempting to build an internal plan of the surrounding world, the robot responds in a pragmatic manner to individual stimuli as they occur. Nothing inside Genghis represents an intention to follow anything or a strategy by which it could be reached. The appearance of motivation and intent arises simply from the blind, autonomous activity of a collection of microcircuits.

Another robot from the MIT stable, Kismet, is a humanoid head that, like Lucy, is designed to interact at a personal level with humans. Pictures show an assemblage of naked mechanical, optical, and electronic parts rendered waiflike by a pair of huge blue eyes capped with silver-foil eyelids, expressive eyebrows (bits of carpet), and a mobile mouth with lips made of red rubber tubing. Cameras sit behind the humanlike eyeballs and are hidden where Kismet's nose would be. Microphones serve as the ears, and the head contains a gyroscope. Servomotors let Kismet move her neck freely and allow her eyes to roam in any direction. She generates facial expressions, opens and closes her jaws, and moves her lips, eyebrows, and ears. Kismet is capable of a limited set of vocalizations: not speech as such but rather like an infant's babbling, keyed to different moods. Her "mental" processes are provided not by a "brain" in the customary location but by a cluster of off-board computers. These process visual and auditory signals in real time, control visual attention and eye and neck movements, and generate expressive speech sounds. Her system architecture structure, displayed on the Web site of Cynthia Breazeal, her creator and minder, is organized into functional blocks. These "orient eyes and head," and perform "feature extraction," "attention," and "vocal acts," as well as seemingly higher level processes such as "emotion" and "motivation."

Video clips of Kismet interacting with a human show her gazing with interest at the face before her. As her interlocutor speaks, Kismet appears to listen intently and react in animated fashion to the tenor if not the meaning of the words. She bats her silvery eyelids seductively, leans forward with ears cocked as if to catch every last word, stares in openmouthed surprise or backs off as though threatened. When the human has finished speaking, Kismet carries the tempo of the conversation with a seemingly meaningful babble that echoes the emotional timbre. Left on her own, Kismet expresses boredom and searches the vicinity for sources of stimulation, focusing on anything that catches her interest, especially a nearby human. Continuous contact with a human seems to develop Kismet's abilities. People interacting with her are said to adopt the role of a teacher or caretaker, guiding her development as they might an infant's. It all seems, to an outside observer, convincingly humanlike.

A more pragmatic application of robots can be found in the sphere of biological research, especially the study of animal behavior. Insects are often the objects of this work, with modeling in machine form of visual perception in moths and flies, location of species-specific songs by female crickets, the climbing of cockroaches, and the escape response of crickets to a sudden puff of wind (as though from an impending predator). Guy Theraulaz, a French biologist working in Toulouse, has shown that important features of nest building by colonial insects such as ants and termites are the result of simple rules. If each ant follows such rules as IF GRAINS OF SAND ARE SCARCE, PICK ONE UP AND CARRY and IF GRAINS OF SAND ARE PLENTIFUL, DROP THE ONE YOU ARE CARRYING, and also responds to such spatial cues as a diffusing source of pheromone, then large-scale structures can result. The elaborate construction of the superbly ventilated termite nests by large numbers of autonomously moving units is an example of what has been termed "swarm intelligence."

Vertebrates too have been modeled. An international team of researchers led by Auke Ijspeert in Lausanne used a robotic salamander as a test animal to analyze the transition from swimming to crawling. The spinal cord of amphibians contains chains of oscillatory centers—groups of nerve cells that produce rhythmic activity and cause local muscle to contract. One set of oscillators produces S-shaped waves of muscle contractions that propagate from head to tail and are responsible for swimming. A second set, associated with the four limbs, forces the oscillations into standing waves—that is, ones that go side-to-side but do not travel—thereby producing diagonally opposed limb movements.

To test their ideas of how the two chains are coupled, Ijspeert and his colleagues built an eighty-five-centimeter-long robot they called *Salamander robotica,* equipped with four rotating legs and six movable joints along its body. Oscillations in the centers were driven at different rates (in the real animal these signals come from the midbrain), and electric motors installed along the body responded. Consistent with their mathematical models, the researchers found that low levels of activity from the "midbrain" produced the relatively slow motion of walking. As the level of stimulation increased, there was an abrupt transition: the walking gait ceased and a more rapid swimming motion ensued. The close similarity of the robotic responses to those of the real animal not only supported the models of neural control but also suggested how terrestrial locomotion might have arisen during vertebrate evolution. Walking could have arisen, they suggest, by modifying and adding layers to a more primitive neural circuit for swimming rather than by evolving a completely new circuit.

What can we say about *Salamander robotica*? Put her into a small tank of water and she swims from side to side, apparently seeking to escape. Place her in the shallow waters of Lake Geneva and she swims to shore and waddles onto the beach in a convincingly reptilian manner. According to Ijspeert and his colleagues, the pattern of electrical signals that drive *Salamander robotica*'s movements is closely similar to those that pass down the spinal cord of a living salamander. But there the resemblance ends. The signals travel along metal wires, shunted by electric switches and activating electric motors. This robot has no eyes, no ears, no sense of smell or touch. The appendages that serve as its limbs are simple plastic pod-shaped extensions lacking skin, blood, nerves, and muscles. They cannot sense roughness, temperature, movement; they are unable to grow and repair themselves. These surrogate organisms are just models in the true sense of the word: simplified representations designed to facilitate calculations and predictions.

The creators of Lucy, Genghis, and Kismet have a more ambivalent attitude. In describing the software that went into Genghis, Rodney Brooks calls it a collection of simple autonomous computational units that, "when placed together and in the context of the physical robot, brought that inanimate object to life. It transcended the boundary between living and nonliving." Remarking that Genghis appeared to have intentions where none were represented internally, Brooks wonders whether the intentions and motivations of real animals might similarly emerge as distributed functions of autonomous units. The robot Kismet, far more advanced computationally, can engage a human in what seems to be, from the outside, a meaningful dialogue. Breazeal's Web site describes Kismet as an "altricial system"—an entity, that is, born or hatched in an immature, essentially helpless condition—"similar in spirit to a human infant," and declares the aim of her project to teach the robot social and communication skills. But no one has ever said that Kismet actually understands what is going on: there is no evidence that she recognizes faces, parses speech, or actually experiences the emotions she simulates. So I am uncertain about her raison d'être. Is it to re-create in a machine humanlike motivations and emotions? Or is Kismet a vehicle to study human responses, a psychological tool to learn how humans subtly mold their speech and movements when confronted by a robot?

Elsewhere, Brooks expresses doubts. Clever though Genghis is, he admits, it is still woefully inadequate when compared with the performance of a real insect. More generally, Brooks comes to the conclusion that despite all the huge advances he and others have made, we are still not able to build machines with the properties of living organisms. He concludes that there is something missing: a kind of mathematics or computation as yet undefined.

> Perhaps we have all missed some organizing principle of biological systems, or some general truth about them? Perhaps there is a way of looking at biological systems which will illuminate an inherent necessity in some aspect of the interactions of their parts that is completely missing from our artificial systems. This could be interpreted to be an elixir of life or some such, but I am not suggesting that we need to go outside the current realms of mathematics, physics, chemistry, or biochemistry. Rather I am suggesting that perhaps at this point we simply do not get it, and that there is some fundamental change necessary in our thinking in order that we might build artificial systems that have the levels of intelligence, emotional interactions, long-term stability and autonomy, and general robustness that we might expect of biological systems. In deference to the elixir metaphor, I prefer to think that perhaps we are currently missing the juice of life.

This, as far as I can see, ten years later, is still where we are. For all the walking, talking, housekeeping, car-building, child-entertaining, elderly-caring marvels of present-day robots; for all their amazing, superhuman ability to crunch numbers, play chess, store data, analyze sequences, and display graphics . . . they are still not like us. Superb specialists, able to do one thing supremely well, robots and their computer brains lack the adaptability and the self-regenerating generalist abilities of living organisms. Above all, they lack something possessed by even the simplest organism: the capacity of independent survival in the real world.

So what might this missing juice be? The most conspicuous difference, implied by the word *juice,* is that living creatures are made of soft, wet, carbon-based materials as opposed to the hard, dry silicon and metal of computers and robots. Living creatures make use of a different, and unique, form of chemistry that enables them to regenerate by reproducing their bodies using commonly available starting materials. That certainly is part of it: a heart pacemaker or a prosthetic limb may serve a particular human function as well as, or even in some respects better than, its human equivalent. But these inanimate structures do not grow and repair themselves. They do not adapt to changing circumstances in the same way as the natural structures. Any attempt to match organisms in this regard would call for machines able to mine ores, extract metals and silicon, refine, mill, draw parts, have the equivalent of a factory assembly line plus a hi-tech facility for the production and assembly of microchips, all driven by solar power or other readily accessible form of power—patently futile.

And then there is the question of chemistry. What would it take to reproduce in a human-made device the myriad chemical reactions each living cell performs every second of its existence? It is true that the broad map of biochemical events in a cell has been drawn and that many of its individual steps and processes can be performed outside a living cell. Sugars and amino acids can be made and modified; large molecules such as proteins, RNA, and DNA can be synthesized from small molecule precursors. But the only feasible way to reproduce entire processes such as metabolism, oxidative phosphorylation, or protein synthesis in a test tube is to use molecules already made by cells. The achievements of contemporary synthetic biology rest on the manipulation of existing cells and biological molecules. Not only are biologists incapable at present of manufacturing the enzymes, membranes, and organelles needed for these processes; they still do not fully understand how they work or are put together.

But you may say that these are side issues. You are not interested in machines that self-replicate or metabolize sugars. Why should you be? You could in principle manufacture as many robots as you wish and give them all the energy they need. No, the crucial question is whether—you might say when—humans will build machines made of steel and silicon that reproduce their emotions and cognition. Our culture is imbued with fictional stories about sentient robots, such as those created by Isaac Asimov in the early 1940s: "RB-34—otherwise known as Herbie—lifted the three heavy books from her arms and opened to the title page of one: 'Hm-m-m! Theory of Hyperatomics.' He mumbled inarticulately to himself as he flipped the pages and then spoke with an abstracted air, 'Sit down, Dr Calvin! This will take me a few minutes.'"

How close are we to actually constructing Herbie? We know that brains and computers both operate by electricity, both specified by billions of connections. What each does or feels is a matter of a pattern of activity— what else? In the world of science fiction, the transition from one to the other is trivial. You simply have to copy the wiring diagram from networks of nerve axons in the human brain to solid-state circuits in a computer. What is to stop you?

TWELVE

The Juice

Estimates of the number of nerve cells in an adult human brain usually weigh in at around one hundred billion. That means 100,000,000,000 self-contained entities, each a molecular metropolis in its own right. (The number of supporting glial cells, which some experts believe contribute to learning and mentation, is five to ten times higher.) Nerve cells, or neurons, have a cell body containing the nucleus and numerous long, thin extensions. Usually, one long axon conducts signals away from the cell body toward distant target cells, and many shorter branching dendrites act like antennae or whiskers, collecting signals from other cells. The axon commonly divides at its far end into many branches, each ending in a specialized site of contact, or synapse. Each synapse allows the signal to be carried to a neighboring cell—either another neuron or a muscle or gland cell. Dendrites can be enormously branched; in some cases as many as 100,000 inputs are received by a single neuron.

Nerve signals take the form of electrical pulses called action potentials. This is true in every part of the brain and spinal cord and whatever information is being carried. Visual information from the eye, motor commands to a muscle, mental thought processes are all carried in the same way. Action potentials depend on the movements of ions (atoms or small molecules that carry an electrical charge) across cell membranes, driven by pumps and channels made of protein. Because of these ionic movements, the inside of an axon is negative with respect to the surface—there is a voltage across the cell membrane. This voltage is small in electrical terms, less than 0.1 volts. But because it occurs over such a short distance (the thickness of the cell membrane), it can have a large effect on proteins in the membrane. When a cell receives a signal, some of the pumps change their rate of action, and the interior becomes transiently positive. A flow of current is triggered that continues down the axon as a wave. Action potentials travel much more slowly than electricity in a copper wire: fat axons may conduct signals at one hundred meters per second, but thin axons and dendrites are slower.

Synapses consist of two parallel plates of membrane separated by a gap that, although small, acts as an electrical barrier. The action potential cannot cross this gap: instead, it triggers release of a substance known as a neurotransmitter—commonly the amino acid glutamate. Glutamate is packed in high concentration within small spherical membrane-bound organelles called vesicles, located at the terminus of the sending axon. When the axon fires, these vesicles move to the membrane on one side of the synapse (the presynaptic side), make contact, open, and release their contents. Diffusing across the cleft, the glutamate triggers a second electrical signal—this time in the receiving cell.

It seems a bizarre arrangement: why not just have an electrical connection? The answer appears to be, first, that a synapse is a one-way route: it allows signals to pass only from the presynaptic axon to the target cell, thereby stamping a directional arrow onto the flow of information. Second, the apparatus of the synapse—the vesicles, their release, the diffusion and capture of glutamate—can all be changed, thereby allowing control of the connection strength. As the part of the nervous system that is most easily modifiable, the synapse has the biggest part to play in learning.

In her 2004 book *Machines Who Think,* Pamela McCorduck traces the history of computers. From the start, she says, there was a notion that they are in some way similar to living brains. The idea was already there, tacitly, in the 1943 paper by McCulloch and Pitts, who represented nerve cells by computer-based logical elements. It was also implied by the work of the English mathematician Alan Turing on machine intelligence. But according to McCorduck, the first to draw explicit parallels was John von Neumann. In a proposal to build a bigger, faster computer, written in 1945, von Neumann made direct comparisons between the parts of the computer and the human nervous systems, employing terms such as *memory* and *control* and referring to sensory, motor, and associative neurons. His ideas were revolutionary at the time. Today they are commonplace, part of our worldview.

Paradoxically, we are now so familiar with electronic computers that analogies are more often made in the opposite direction: brain functions are commonly interpreted in terms of computers. This aspect of wetware exists at a higher level than the computational processes described elsewhere in this book, but there are interesting parallels between the two. Thus the logical processing that at the molecular level is accomplished by protein molecules is, in the brain, performed by neurons. Signals inside a cell are carried by diffusion, but in a brain they move as electrical signals. The relationship between these two levels is even more interesting when we consider their mutual dependence. For the computations of the brain arise—are built step by step—from those in individual cells. Protein molecules with their switchlike properties are essential ingredients for action potentials in axons. Proteins transmit signals across synapses; they are the microcircuits that store memories.

A typical nerve cell has perhaps 1,000 synapses on its surface (some have many more than this), each a point of connection with another nerve cell. So if I multiply the number of synapses by the number of nerve cells—1,000 times 100 billion—I arrive at 100 trillion (10^{14}) as the number of synapses in a human brain. If each synapse can be either on or off—the simplest possibility—then each will contribute one bit of information. The storage capacity of an adult human brain then becomes around 10^{14} bits, or 12.5 terabytes in computer terminology. For comparison, this is about the quantity of information in seventeen thousand CD-ROM disks, or the capacity of ten modestly sized academic libraries.

How reasonable is the one-bit, one-synapse assumption? How much information do you expect a human brain to carry? Impossible to measure, of course, but there have been guesses. Probably the upper record was an estimate by von Neumann, who foresaw many developments in computer science in his pioneering book *The Computer and the Brain* (1958). What if, he asked, the brain were to keep a complete record of every electrical signal generated in the course of sixty years? His calculations led him to a huge number corresponding to something like 30 billion bits of information storage per neuron—perhaps 100 million bits for every synapse. Evidently, this is a wild overestimate. No one seriously believes that each action potential in every nerve axon is recorded over a lifetime.

At the other extreme, there are practical estimates of how much we actually store and recall. In 1986 Thomas Landauer at Bell Communications Research analyzed human memory as if it were a telephone line carrying information (he was working at the same institute that had supported Claude Shannon in his pioneering work on information theory). Landauer tested 204 fellow employees, local homemakers, and Princeton undergraduates on two tasks: one involving concentrated reading and the other measuring the recognition of pictures and other material. After delays ranging from a few minutes to many hours, subjects were tested to see how much they remembered. They were asked to fill in words randomly deleted from a previously viewed text, or to identify pictures they had seen before.

The conclusion of this study was that humans store information in their long-term memory at a remarkably constant rate of a few bits per second. Over a lifespan of sixty years, this corresponds to a storage capacity of around 4 billion bits—considerably less than the total number of nerve cells available. It was clear to Landauer that the quantity of information he measured was a gross underestimate of total capacity. He was assessing consciously accessible memory, so his tests ignored the neural capacity used up in internal maintenance, such as the control of heartbeat, breathing, swallowing, and the chemical balance of the digestive organs. They made no account of bookkeeping, database management, and the control of noise, unreliability, and damage. All of these functions require neuronal circuitry and, no doubt, modifiable synapses.

Two estimates of brain storage capacity place the information content of a synapse somewhere between one hundred million bits and considerably less than one bit—a gap the size of the Grand Canyon. Can't we do better than this? Another tack, more in keeping with the perspective of this book, is to consider the structure of a typical synapse. How large is it? How many molecules does it contain? How different are synapses one from the other? How do they change over time and with the storage of memories?

Microscopic examination of a section of cortex, the major and most distinctive part of the mammalian brain, reveals one type of nerve cell that outnumbers all others. Known as pyramidal cells, from the shape of their cell body, these were first detected in thin sections of chemically hardened brain. In the late nineteenth century, the Italian physician and scientist Camillo Golgi discovered a way to stain such sections with silver compounds. For reasons still not fully understood, the Golgi method picks out a few individual nerve cells from a population and stains their entire form. Every detail of the cell's shape is drawn in deepest black against a translucent, slightly creamy background. Pyramidal cells from mammalian brains appear like fir trees on the side of a snow-clad mountain. Each has a single dendrite that rises toward the cortical surface, where it produces side branches and numerous shorter dendrites that fan out like roots from its base. A single long axon heads south, departing for distant synaptic destinations. The number and extent of pyramidal cell dendrites and their abundance in regions of mental function led the famous histologist Santiago Ramón y Cajal to call them, in characteristically vivid fashion, the "cellules psychique"—the psychological cells. More than any other type of nerve cell, Cajal suggested, pyramidal cells capture the deep association among cell shape, connectivity, and the higher functions of the nervous system.

Along most of their length, pyramidal dendrites are covered with minute thornlike projections, or spines, each a few microns long. There are more than ten thousand on a typical human pyramidal cell. Each is the visible evidence of a synapse—the site of contact between an axon coming from another cell. Action potentials in the incoming axon enter the spine and release neurotransmitter. On the other side of the synapse, the pyramidal cell detects the diffusing neurotransmitter and generates an electrical signal. This may be either excitatory—tending to produce action potentials in the receiving cell—or inhibitory. But because they are so numerous, the effect of any individual spine is rather small, and the net result on the cell as a whole is a weighted sum of their individual actions. A pyramidal cell seems to be a computational unit whose output, or firing rate, is determined by the strengths and positions of all the synapses on its surface.

To get a closer look at a synapse you need to view thin sections of brain at higher magnification in an electron microscope. Now the beautiful simplicity of silver staining is lost and the tissue is revealed as an inelegant jumble of cellular profiles that are contiguous and confusing. With experience and patience, however, axons and synapses can be located in this hodgepodge. Synapses look like miniature pomegranates full of tiny membrane vesicles, each a tiny container full of the amino acid glutamate, the neurotransmitter, ready to be released from the axon during electrical excitation. Next to each axonal terminal is the dendritic spine—the receiving end of the synapse. The distance is tiny—about twenty nanometers—which can be compared

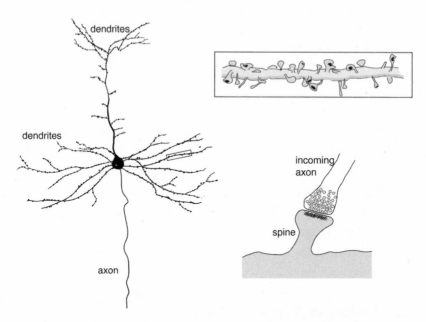

FIGURE 12.1. Synapses. A typical pyramidal nerve cell from a mammalian brain is shown at left as it appears in silver-stained preparations. Numerous branching dendrites are covered with thornlike projections or spines (upper right), each the site of connection, or synapse, with another nerve cell. Details of a single synapse are shown at lower right.

to the thickness of the cell membrane (about seven nanometers) or the size of a typical globular protein such as actin (five nanometers).

It's curious to realize, in light of previous chapters, that a dendritic spine is about the same size as a bacterium. Moreover, the membrane of each spine contains a flat roughly discoid plate of proteins, the post-synaptic density, similar in size to the cluster of chemotaxis receptors. Both structures are also concerned with the chemical detection of small molecules outside the cell, but there the similarity ends.

The postsynaptic density of a dendritic spine is far more compli-cated. It contains the apparatus for detecting and processing pulses of glutamate and includes more than two hundred kinds of proteins. Their spatial arrangement is not understood in detail but is certainly elaborate and probably capable of extensive variation. Combinations of high-resolution microscopy and molecular probes hint at a grid or ma-trix of fibrous protein just beneath the plasma membrane. Attached to this scaffolding and projecting into the cytoplasmic region of the spine is an irregular superstructure (or substructure, depending on your point of view) built from signaling proteins. A conspicuous feature is a collection of a dozen or so cylindrical pillars made of a remarkable pro-tein called calmodulin-activated protein kinase, or CamII kinase.

The first and most basic requirement for the synapse is to act as a switch. The presynaptic portion—belonging to the axon—first receives the action potential. When this happens, the small voltage difference across the plasma membrane, typically sixty mV, with the inside nega-tive, is transiently reduced. If this change is sufficiently great, channels in the membrane open, and calcium ions flood into the axon terminus. The increase in calcium ions triggers the movement of vesicles to the presynaptic membrane, where they fuse, open, and spill their content of glutamate into the synaptic cleft. Crossing the minute gap in a matter of milliseconds, the released glutamate molecules bind to specific re-ceptors in the membrane of the receiving cell.

Binding to the receptors on the postsynaptic side, the glutamate causes activation—that is, a change in molecular shape of the receptors—and triggers intracellular signaling in the dendritic spine. The plasma membrane of the receiving cell also becomes depolarized, creating an electrical signal that contributes to the firing of the cell. The activation of the receptors also sets in train cascades of signaling reactions within the postsynaptic cell. The nature and extent of these signals, and the changes they cause, are influenced by the size and frequency of the glutamate release.

Not every action potential arriving at the synapse has the same effect. Like waves on a seashore, voltage fluctuations arrive with similar but nonidentical shapes; their cumulative effect depends on how close together they are. A burst of action potentials close together causes a relatively large depolarization of the membrane, leading to a greater release of glutamate into the synapse. The effects at the receiving end will be correspondingly greater: a bigger swing in the membrane voltage of the receiving cell, an increased chance of triggering an action potential. There is also evidence that the synapse has a greater chance of becoming stronger if the receiving cell happens to be firing at the same time—a crucial element in many theories of learning.

Perhaps the single most important feature of the brains of higher animals is their ability to learn. The neuronal circuitry of a newborn baby, made by the autonomous processes of cell development and differentiation within the womb, offers a clean slate to the world. After birth, sensations and experiences flood in from the surroundings, shaping the connections between nerve cells and hence the responses of the growing child. Some new nerve cells are born during this period, and many others die, being superfluous to requirements. But the most widespread and pervasive changes are to synapses. These continue to be added, lost, or changed in strength throughout life.

Evidently changes in synapses alter the connections between nerve cells—change the neural circuit. But how can the cells in the brain know which of zillions of synapses to alter, and by how much? The Canadian psychologist Donald Hebb offered what is still considered the most plausible strategy in his 1949 book *Organization of Behavior*. "When an axon of cell A is near enough to excite a cell B and repeatedly or persistently takes part in firing it, some growth process or metabolic change takes place in one or both cells such that A's efficiency, as one of the cells firing B, is increased."

Note that Hebb's prescription does not imply action at a distance; he was most emphatic about that. His two cells, A and B, were already linked by a synapse, presumably having become connected during the growth of nerve processes during development. Something like 50 percent of the synapses in a newborn brain are "silent"—having no detectable function. So you can, if you wish, imagine previous unused synapses becoming larger and gaining a voice because of the coincident firing of two nerve cells already in contact.

Think of a dog in a Pavlovian experiment, being trained to salivate when a bell rings. Many nerve cells will participate in this new reflex, but the eventual circuit must be built step by step by the reinforcement of synapses between contiguous cells. Suppose, for example, that nerve cell A fires whenever the bell rings and nerve cell B fires when the dog salivates. Suppose, moreover, that, for reasons completely external to the dog's nervous system, whenever the bell rings, food is produced. You and I know, of course, that the experimenter causes this temporal coincidence. But the dog's brain knows only that cell A and cell B repeatedly fire at about the same time. No matter—provided it follows the rule prescribed by Donald Hebb (or something like it—there are numerous possible refinements of the basic mechanism), then the connections between A and B will become stronger. The activity becomes reinforced, and the dog salivates on hearing the bell, even in the absence of food.

You can go further. The basic machinery of synaptic transmission provides a plausible theory of how such a Hebbian synapse might work. Assume that when the bell rings, cell A fires an action potential. The wave of electrical excitation travels down its axon and arrives at the synapse with cell B. What you now need is for this synapse to DETECT THAT CELL A AND CELL B ARE ACTIVE AT THE SAME TIME and, if this is the case, MAKE THE SYNAPSE LARGER AND MORE EFFECTIVE.

How might it accomplish these two tasks? Well, to begin with coincidence detection, it seems logical that the receptors of the synapse between cell A and cell B will be crucial. You know that they bind to glutamate released into the synaptic cleft, thereby registering the fact that cell A has fired. You also know that they are embedded in the membrane of cell B, so they are ideally positioned to read its membrane voltage.

And as you have probably guessed, the receptor does indeed meet these specifications. At least, one specific kind of glutamate receptor does (synapses have several different types). This protein generates a signal only when the synapse fires, releasing glutamate, *and* the receiving cell also carries an electrical signal. A glutamate signal by itself is not enough.

Under resting conditions the mouth of the receptor is blocked by magnesium ions. These positively charged ions, abundant in the surrounding fluid, are attracted to the membrane by the voltage across it (the inside of the membrane is negative). But if the membrane voltage is reduced because the receiving nerve cell fires, the blocking magnesium ion is released. A receptor bound to glutamate now becomes active and triggers a cascade of events, starting with an influx of calcium ions into the receiving cell. These flood into the synaptic spine and trigger changes leading to a reinforcement of the synapse.

But think what I have just said. The glutamate receptor—a protein inserted in the membrane—detects two features of its surroundings. It detects both the amino acid glutamate released from cell A and also a low membrane voltage in cell B. Only when both conditions are present does the receptor become active and generate a signal. So in effect this protein is performing a simple computation—a Boolean AND function. It is yet another protein switch, like the others I described in Chapter 3. Once again, a protein molecule links two distinct functions in the cell.

The glutamate receptor is slightly different from the earlier examples in that one of its two inputs is electrical in nature. But it must, nevertheless, have evolved sets of amino acids to detect this voltage change. It has to hold or release the blocking magnesium ion depending on the size of the voltage across the membrane. The voltage change is one input signal, and the glutamate released into the synaptic cleft is the other. The output signals the simultaneous firing of two contiguous cells. The coincidence of these two events is crucial to the establishment of functional links in the brains of higher animals—the first requirement for learning to occur.

Again there is resonance between single cells and higher organisms. Protein switches linking two functions are at the heart of the molecular circuitry of a bacterium or an amoeba. Now you see that in higher animals, including humans, the association of stimulus and response during learning has, deep down, the same molecular basis. Learning and, by extension, the mental association of ideas and our most intimate thoughts are built from subtle changes in proteins. As with hunger and respiration, a fundamental molecular process in individual cells reemerges at a much higher level in a multicellular organism.

I still have to explain the second stage of learning. How does the firing of the receptors cause the synapse to become stronger and more effective? How does a fleeting event—the temporal coincidence between cell A and cell B, bell and saliva—become a long-lived memory? One of the most plausible answers to this question involves a protein already mentioned, CamII kinase. This remarkable enzyme, abundant in the dendritic spine of the receiving cell, becomes activated by the sudden influx of calcium ions. It then adds phosphate groups to target proteins in its vicinity, thereby triggering a cascade of biochemical changes, similar in principle to those involved in glucose metabolism in a liver cell or chemotaxis in a bacterium.

But CamII kinase is set apart by the subtle, reflexive, nature of its activity. For among its many targets this enzyme also adds phosphate groups to itself. Moreover, if it acquires sufficient phosphates in this way, it becomes irreversibly active, beyond the brief rise and fall of calcium ions. In other words, the enzyme acts like a toggle switch that, once thrown, remains on forever . . . the epitome of a memory molecule. A synapse containing persistently active CamII kinase molecules then becomes more sensitive and gives a stronger response. It acquires new proteins, such as a second kind of glutamate receptor that fires with every puff of glutamate. The synapse becomes larger and more convoluted in shape.

You can see why investigators in the field are excited about CamII kinase. It becomes active (adds phosphates) only in a synapse that has experienced a large rise in intracellular calcium. This happens only when both partners of the synapse, cell A and cell B, are active at the same time. From that point on, provided the calcium signal is big enough, the CamII kinase stays switched on. Any phosphate groups that are by chance lost through the action of a phosphatase are rapidly replaced by self-remediation, one subunit of the enzyme acting on its next-door neighbor.

Amazingly, the active state can be preserved for months, even years—well beyond the natural life of the proteins. The fabric of my body is subject to a continual process of chemical damage by processes such as oxidation and radiation. Old, damaged proteins are removed, degraded, and replaced by newly made molecules at a surprisingly high rate. Most proteins in an adult turn over every two to three weeks, some do it in less than twenty-four hours. But if this is so, how can I store long-term memories? I can remember events that occurred half a century ago; what protein-based structure could last that long?

A possible answer to this conundrum comes from the cooperative nature of the CamII kinase machine. A subunit lost through the inevitable process of decay and turnover will be replaced by a new one, diffusing from the site of synthesis and wriggling into place like a puppy rejoining its litter. Initially this newcomer will be inactive because all proteins are made in a naïve, unmodified state. But if the protein complex as a whole is currently active, then the newcomer will quickly adopt the same state, receiving phosphate groups from its neighbors. In this way, the tiny fragment of memory encoded by that CamII kinase is retained: the memory will outlast the molecules of which it is composed.

There is strong support for this picture. Not only does Cam II kinase have a suitable location in a synapse and appropriate activity. Mouse mutants with defective CamII kinase or none at all seem unable to store memories, especially spatial memories. Particularly striking is a mutant enzyme unable to add a phosphate group at a particular location on its surface. A mouse carrying this mutation and placed in a water-filled tank (a commonly used test for learning) repeatedly fails to find a submerged platform to stand on: it behaves exactly as an animal that has lost its ability to store spatial memories.

So perhaps I have found what I was looking for? Could it be that Cam II kinase in each synapse is analogous to a transistor in a printed circuit . . . contains an elementary bit of information?

Well, since it exists in two well-defined states, active and inactive, a single CamII kinase molecule could in principle serve as a Boolean switch (although it would be noisy and imperfect). But then, if I view the synapse as a whole, I realize it has a far richer structure. There is much more to a synapse than a single Boolean switch.

There is, to begin with, not one CamII kinase molecule in each synapse but a dozen or so. Acting independently, each could in principle act as a switch—store a separate bit of information, if you will. But since they are embedded in the same cluster of proteins and associated with other signaling proteins, they might work together. They could, for example, cooperate to amplify signals and make them easier to detect and to measure (as I described for bacterial chemotaxis). They might increase reliability and longevity.

Brain scientists admit that they currently have little notion what computations can be performed by a single synapse. Action potentials arrive at the synapse side, and the membrane voltage of the receiving dendrite on the other side of the synapse rises and falls. But there is a world of possible variation in these fluctuations. Action potentials might come at unpredictable times, spaced one from the other with subtly different heights and waveforms. The synapse they enter is a complicated space occupied by vesicles, filaments, and membranes. Hundreds of proteins are there, associated with structures or freely diffusing in solution. Each of these proteins is, as we have seen, potentially able to exist in multiple different chemical states and to interact in numerous ways with its neighbors. The size, position, and precise timing of glutamate release are likely sensitive to the precise molecular landscape. And the receiving end of the synapse in the dendritic spine is if anything richer in molecular diversity.

Timing is also crucial. Rhythmic processes are ubiquitous in the brain, as revealed by the electroencephalograph recording of brain waves. Although opinions on the importance of these rhythms differ, some experts believe that oscillations are crucial for mental functions including learning. One theory holds, for example, that neurons that fire in synchrony are bound together in a semantic sense—associated with the same concept. So it could be important for a synapse to be able to detect time lags and frequencies. Even a single CamII kinase can discriminate among frequencies of calcium change. It is not difficult to imagine molecular circuitry that would lead a synapse to become stronger if it received impulses at a certain frequency.

Each nerve cell has thousands of synapses. The large pyramidal cells might have tens of thousands. So a single nerve cell in your brain is very complicated. In fact, by any practicable measure, it must be just as complicated as any free-living cell. Each of your pyramidal cells has as many different kinds of protein molecules arranged in just as many intricate structures as any amoeba or stentor—no, more. Human cells have more genes and are also richer in mechanisms of gene variation than organisms lower in the evolutionary scale. In fact, for reasons you might guess at, alternative RNA splicing and protein modification are most exuberantly expressed in mammalian brains.

So you have every reason to expect pyramidal cells to be capable of highly subtle individual responses and behaviors. Recall how single-celled organisms hunt for food, respond to light, and react to sound (at least vibrations) and temperature. Think of the elaborate avoidance responses stentor displays when it encounters a noxious stimulus, how it twists, cowers, and eventually breaks free, how it swims away in search of a new location and sets up a new home, how it behaves as though it were a complete organism.

Imprisoned in the crowded confines of the cortex, your pyramidal cells have no opportunity to escape. But there is nothing to stop them moving in place. They can grow and shrink, extend or retract their dendrites and branches of their axons. Their synaptic spines can be every bit as plastic in form, just as dynamic as an amoeba's pseudopodia. The movements of a nerve cell in the brain may be of shorter duration and harder to observe than those of a free-living cell, but why should they be less well informed? The milieu of such a cell is a rich and ever-changing broth of ions and neurotransmitters. There has been every opportunity for nerve cells evolving over millennia to learn how to extract information from this chemical soup—to recognize important changes. Having at hand the universal machinery of shape change and movement, why not harness this machinery to the task of processing electrical signals? With thousands of synapses on its surface forming contact with perhaps hundreds of other cells, a pyramidal cell is in a key position to perform a sophisticated assessment.

So what can a single nerve cell do? When it fires, what message does this convey? Questions such as these have been asked for as long as neurophysiologists have pushed electrodes into animals. In his 1928 book *The Basis of Sensation,* Lord Adrian relates:

> I had arranged electrodes on the optic nerve of a toad in connexion with some experiments on the retina. The room was nearly dark and I was puzzled to hear repeated noises in the loudspeaker attached to the amplifier, noises indicating that a great deal of impulse activity was going on. It was not until I compared the noises with my own movements around the room that I realized I was in the field of vision of the toad's eye and that it was signalling what I was doing.

Adrian's student Horace Barlow, now emeritus professor in Cambridge, was the first to link neural activity directly with perception. His central proposition was that single neurons code for perceptually significant events (rather as, I suggest, single proteins come to represent important features of a cell's environment).

Around 1969, in a course he was giving at MIT, the flamboyant neurophysiologist Jerome Lettvin coined the term *grandmother cell*—the hypothetical nerve cell that fires only when presented with an image of one's grandmother. Lettvin's remark was tongue in cheek, deliberately exaggerating the trend of discovery by him and others of nerve cells' response to complicated stimuli. The Polish neurophysiologist Jerzy Konorski, in his 1967 book *Integrative Activity of the Brain*, had made a similar proposal independently, using the more prosaic term *gnostic units*.

Konorski's ideas were influenced by observations of the effects of local damage to the brain. Memories that were lost often seemed to belong to a definite category, such as vehicles or animals. That cells could respond to highly specific stimuli, purely theoretical at the time, is now taken quite seriously. There have even been reports of cells that respond selectively to the faces of well-known celebrities. The issue is still debated, but everyone seems to agree that individual cells in the mammalian cortex have extremely sophisticated processing capabilities.

A neuronal synapse reveals most clearly the distinction between living and nonliving computers. Since it carries information from one nerve cell to another, you might be tempted to represent it as a single transistor in a printed circuit—as a single bit of information. But this would be to miss the point. Far from performing in a rigid, stereotypical, predictable fashion, synapses are richly, almost infinitely, variable in their input-output relationships.

A synapse, about the same size as a bacterium, contains a plethora of proteins that are continually degraded and renewed. Through an unending cycle of molecular death and rebirth, synapses morph their composition and structure to recent events. Fluctuations in membrane voltage, release and uptake of neurotransmitters and ions, frequency and shape of action potentials are all relevant. These transient effects can regulate local processes such as protein modification, assembly of protein complexes, and intracellular movements. Local physiology can even influence events in the distant cell nucleus such as gene expression and alternative RNA splicing. By embedding a record of its past in its molecular structure, each synapse becomes distinct, unique. Its response to patterns of activity will be distinct from that of any other synapse. Most of this variation will be meaningless, just chemical noise. But a significant fraction of the multiple forms will be significant and will mean something in terms of the outside world.

The word *information* comes to mind here. A tricky term—another of those Janus-faced words like *memory,* which mean one thing to you and me and something else to the experts. From a technical standpoint, using an analysis developed for the transmission of messages along telegraph wires, a synapse that either fires or does not contributes one bit to the information transmission capabilities of a neural circuit. But such a measure ignores the representational capacity of a synapse, the number of different states that it can store.

You might hear here an echo of other processes discussed in this book. Recall how during evolution the sequence and structure of proteins morph in response to environmental pressures. Remember how antibody molecules are shaped by exposure to antigens in the immune system. A synapse is similarly protean, instructed by the firing of other synapses on the same cell, by axons that connect with that cell, and ultimately by the environment of the organism. Perhaps a better word to use here is *knowledge.* By capturing a picture of their surroundings in molecular terms, biological systems acquire knowledge of the world in a way no other chemical or physical system can.

THIRTEEN

Amoeba Redux

I began this book with the movements of single cells, and perhaps that will be a good topic with which to finish. What can I now say about the predation of amoebae or the avoidance of stentor, given a better notion of how they work? How do these single-cell organisms operate in an integrated, apparently motivated fashion? Where does their sensibility come from? What can I say about their self-knowledge?

The starting point is a cell stuffed full of molecules, especially proteins. Proteins provide the equivalent of muscles, skeleton, digestive system, and lungs. They create networks of communication and logical machines—the substrate for the cell's computations. Where higher organisms have a brain and spinal cord, single cells have networks of interacting proteins.

The processes I've collected in this book under the term *wetware* include most of the chemical reactions inside cells. They include the transformations of small molecules familiar in energy metabolism and the synthetic reactions used to make large molecules; the modifications in structure of proteins by addition of phosphate and methyl groups; the assembly of proteins into large complexes; the turning on and off of genes; the transport of ions and small molecules across membranes; the generation of mechanical force and directed motion.

Exactly where you draw the line between wetware and other functions of the cell is a matter of choice. Some processes are obviously computational in nature: the cascades of reactions that carry signals from the membrane, or that switch on a gene in the nucleus or activate a cell movement, have clear parallels with the nervous system (and with electronic devices). Conversely, it is conventional to regard the enzyme reactions that generate energy or make amino acids as essentially biochemical processes. But they are just as subject to logical control as those of cell signaling. It is just as easy to represent them symbolically as a series of statements in a computer program or, if you prefer, as components of a physical electronic circuit.

The computational processes performed by a cell have features that distinguish them from any human-made machine or electronic device. The most obvious difference being that they are based on chemical reactions in aqueous solution. They require a continual supply of carbon atoms and chemical energy, considerations that do not apply to electronic circuits. Connections in a cell-based computer are made by the physical processes of diffusion and molecular recognition. This renders cell communications comparatively slow and inaccurate, but, as if in compensation, it also makes it possible for large numbers of connections to be made in a minute volume. Cells are not limited by the size of wires or overheating; their energy consumption is minimal. The brain of a fruitfly performs miracles of visual processing and flight control with the consumption of microwatts of power. A supercomputer, by comparison, uses megawatts—a million million times more energy.

A cell operates as a massively parallel system. What a biochemist considers one reaction or the reader of this book might call a single computational step is actually a simultaneous change in many individual molecules—typically thousands in a cell the size of a bacterium or millions for a plant or animal cell. In a well-stirred solution of dissolved molecules you can indeed consider this one reaction the equivalent of a single logical switch in a printed circuit, one transistor. But in the crowded confines of a cell, highly organized in space and time, seemingly identical molecular species take up distinct locations. Precisely where a protein is at any instant of time will dictate its nearest neighbors.

In turn, these influence what will happen to the protein in the next instant. So the single reaction splinters into many slight variants.

Location is crucial. Every type of structure in a cell—membranes, protein filaments, mitochondria, and chromosomes—carries a distinctive chemical label on its surface. And within each structure there are subdivisions. I mentioned in Chapter 8 the histone code, the chemical labels on the surfaces of chromosomes carried by histone proteins that flag particular features of the DNA within. The enzyme that copies regions of DNA into RNA has a tail on which is written, chemically, where the enzyme is in its progression through a gene.

Outside the nucleus, filaments forming the cytoskeleton carry other modifications and different proteins depending on location. A hairlike cilium on a cell surface has a structural core built from a sheaf of tubular protein filaments called microtubules. Not only are these microtubules subtly different from those in the cytoplasm of the cell or associated with chromosomes. They also show regional differences *within* the cilium, with different proteins associated with microtubules near the base and those near the tip. Similarly, cell membranes are far from being mathematically idealized planar surfaces. A classic review by Jonathan Singer and Garth Nicolson published in 1972 described the membrane of a cell as a "fluid mosaic." It was like, they said, a sea of mobile lipid molecules full of floating protein molecules. An appropriate metaphor, since, as we now realize, the floating proteins come together in patches and clusters. They create dynamic archipelagoes, unique molecular islands that form and disperse with the experience and intent of the cell.

What determines this complicated geography? Ultimately, it depends on the now familiar process of molecular recognition. Unceasing movements due to diffusion bring randomly assorted molecules into contact billions of times per second. Any that fit together may make enough weak bonds to stay together, at least for a while. But location is so crucial to efficiency and survival that special mechanisms of delivery exist: courier services that make movements faster and more accurate, albeit with a cost in energy.

Major locations of the cell have something analogous to a postal address. Proteins destined to work in the nucleus, for example, contain distinctive sequences of amino acids rich in positively charged amino acids known as "nuclear import signals." These signatures are recognized by receptors in the nuclear membrane and the molecules conveyed forthwith into the nucleus. Other targeting sequences exist for organelles such as mitochondria and for the many kinds of internal membrane found in eukaryotic cells.

RNA molecules provide perhaps the most remarkable example. Messenger RNA molecules (those that code for a protein) are made in the nucleus. They then move into the cytoplasm, where they are directed to specific locations according to sequences of bases at one end of the molecule. These sequences do not code for amino acids and are not used to make protein; instead, they act as ZIP codes. Their precision can be amazing, some RNAs finding their way to the ends of growing nerve cells, or to the synapses of a large pyramidal cell.

Evidently something more than simple diffusion must be responsible. A molecule the size of RNA would take weeks to diffuse from one end of a large nerve cell to the other. What in fact happens is that motor proteins seize the RNA molecules and then carry them along microtubules. Specific adapter proteins are needed to recognize the RNA, read its destination, and then attach it to a suitable motor protein heading in the desired direction. Once the RNA cargo has reached the correct region of the cell, other proteins cause it to detach from the motor. It begins its appointed task of making a protein.

It is impossible, I think, for us to envision the richness and diversity of cell chemistry. The level of detail is atomic in dimensions but astronomical in variety. Every structure inside a cell is covered with a mosaic of chemical groups, positioned and maintained by the mechanisms just mentioned. Every protein molecule is subtly different, carrying not only the imprint of history, shaped by evolution over millennia, but also an echo of recent events. Given that each protein molecule also interacts with others in multiple ways, the overall picture seems horrifyingly complex. It seems a hopeless task to ever unravel such a web of cause and effect in even a single cell. And yet—in plain contradiction—we do in fact understand a great deal about cells. Libraries full of erudite, fact-filled journals and books testify to the wisdom acquired, especially in the past century. The wonders of modern biotechnology and medicine constitute highly visible evidence of our knowledge. How is this possible?

The solution to this apparent paradox is, in a word, organization. If proteins and other molecules were scattered throughout a cell in a totally random pepper-and-salt fashion, then we would indeed be in trouble. Any attempt to diagram their functional interactions would result in a thicket of lines, impenetrable in both a literal and a figurative sense. But in reality both the locations of molecules and their functions are highly nonrandom. Crucially, they are organized in a hierarchical fashion at multiple levels. Typically, one kind of protein molecule associates with a set of others to make a complex. This complex will have a specific role, as I described in Chapter 5, and may in turn act as a subunit for a larger complex. Large complexes, anchored to intracellular structures such as the chromosome or membranes, create organelles. Organelles come together to make a cell. Nor do I have to stop there, because cells are grouped into tissues, tissues into organs, organs into physiological systems, and systems into a body.

Because of its hierarchical organization, I can dissect a cell into distinct functional units. I can, if I wish, study protein synthesis without worrying about sugar metabolism; measure the oxidization of sugars of mitochondria while ignoring the mitotic spindle and the apparatus of cell division. Proceeding in this fashion I can put together a model of how the cell works as whole. I can test and refine this model by performing specific experiments.

It is fortunate for us that cells are built in this way—but what advantage does it have for them? They are not organized for our benefit. Herbert Simon, the Nobel-winning political scientist who made many contributions to cognitive psychology and computer science, addressed a related philosophical problem in his seminal 1962 essay "The Architecture of Complexity." Most complex systems, he said, are organized in a hierarchical fashion. Simon mentioned molecules built of atoms and subatomic particles; human societies built of nations, tribes, and families; galaxies built of stars and planets. I might add computers and brains because these also are made of many parts linked in a complicated way. Although these systems are as different as they can be in most respects, Simon said they nevertheless have some common properties. These arise because of their organization, which he illustrated in a telling and justifiably famous parable.

There once were two watchmakers, named Hora and Tempus, who manufactured very fine watches. Both of them were highly regarded, and the phones in their workshops rang frequently— new customers were constantly calling them. However, Hora prospered while Tempus became poorer and poorer and finally lost his shop. What was the reason?

The watches the men made consisted of about 1,000 parts each. Tempus had so constructed his that if he had one partly assembled and had to put it down—to answer the phone, say— it immediately fell to pieces and had to be reassembled from the elements. The better customers liked his watches, the more they phoned him, the more difficult it became for him to find enough uninterrupted time to finish a watch.

The watches Hora made were no less complex than those of Tempus. But he had designed them so that he could put together subassemblies of about ten elements each. Ten of these subassemblies, again, could be put together into a larger assembly; and an assembly of ten of the latter subassemblies constituted the whole watch. Hence when Hora had to put down a partly assembled watch in order to answer the phone, he lost only a small part of his work, and he assembled his watches in only a fraction of the man-hours it took Tempus.

A living cell evidently resembles one of Hora's watches. It too is made as a hierarchy of smaller subassemblies, each of which is independently stable. If there are occasional hiccups in the supply of materials or interruptions in the process of biosynthesis, then the process as a whole can keep going. It is hard to comprehend how else cells could be made so rapidly and reliably under such a variety of conditions. The adult human makes a hundred billion red blood cells a day, the vast majority of which function flawlessly. Cells in tissue culture continue to grow and divide despite major perturbations in the supply of nutrients. The numbers of molecules of particular proteins in an animal cell fluctuate wildly, and yet the cell appears unchanged, at least to superficial inspection.

The watchmaker analogy also sheds light on how cells might have evolved. The natural environment is full of discontinuities and step changes, alarms and excursions. Any organism so constituted that it becomes damaged by such interruptions will surely, like Tempus, soon be out of business. A strong selection must exist in favor of stability at all levels of production. Self-sufficient, autonomous complexes and organelles will arise, quite simply, because they outlast the others. They alone survive long exposures to the exigencies of the environment. To this extent they resemble the subassemblies of Hora's watches, except that they are the product of random chance rather than "intelligent design."

There is more to the story. If a living cell were no more than a watch or other piece of machinery, I would have little more to say. I could simply make a list of the parts, describe each one, and indicate how it connects to the whole. The list would be long, but in principle, if you studied it carefully, you would eventually understand how the cell works. Using the plan as a blueprint, you would know the location of every molecule, would be able to predict the precise shape of the cell and what it would be doing at any instant.

But we know that this is not possible. Whatever resemblance a living cell might have to a watch, it is unlike any timepiece made by human hands. This watch is soft and deformable—a Daliesque watch that melts onto its surroundings, one that can adopt more designs than you will find on Macy's Web site. In other words, superimposed on the hierarchical framework of defined components of a cell there is another layer. This second layer is highly flexible and can take on an almost infinite variety of forms, like soft and responsive flesh on a bony skeleton.

The deep question is whether this higher layer in the construction of cells is itself organized. Are there hierarchies, or at least rules, in the protein-modifying, RNA-splicing, gene-regulating processes of a cell? If so, then we have a chance of understanding them. If not, we will never know exactly what a cell will do next. If the detailed chemistry of the cell is simply the outcome of a historical ragbag of ad hoc interactions, then it will be no more predictable than the weather.

I do not have an answer to this question. But two features of cells might be relevant. One is a sense of time, or causation—knowledge of the way that things in the real world follow in a certain sequence. The other is integrity, which enables a cell to distinguish between what belongs to itself and what belongs to the outside world.

The features of cell chemistry written in the sequence of bases in DNA and amino acids in proteins are for the most part long-lasting and stable. They represent the "given" or "prior" knowledge a cell inherits when it is created at cell division. But as the cell feeds, grows, responds, and moves, this static chemistry becomes refined and modified. Thousands of changes occur, such as chemical modifications of proteins or changes in their shape or location. The reversible nature of these modifications, together with the never-ending breakdown and resynthesis of proteins, ensures that they are dynamic and responsive. They represent the cell's most recent experiences.

Why should a cell want to record past events? Is it out of historical interest or nostalgia? Hardly. Events that happened years, days, or even seconds ago are beyond reach; nothing can change them now. But if the cell can use the past to read the future—to gain even the slightest inkling of what will happen next—ah, then, what an advantage it will have! This is why a swimming bacterium stores a record of its environment over the past few seconds—simply to help it predict the future. By comparing the stimulus level in the recent past with present conditions, the bacterium learns whether its environment is getting better or worse. It then has a basis on which to decide whether to continue swimming the same way or try some other direction.

Prediction is easiest when the same event has happened many times before, the most pervasive examples being due to the earth's rotation. Our magnificent sun has stamped its image on not only the energy metabolism of living organisms (through photosynthesis) and their sensory pathways (through vision) but also on their internal rhythms. The daily alternation between light and dark and the longer seasonal changes between winter and summer have shaped our living world.

As I write, the century-old walnut tree outside in my frost-nipped garden looks utterly lifeless. Solid and smooth-boled, its branches a reticulum of dry twigs—it looks massively dead, a wooden statue. Yet I know (and in a sense the tree also knows) that spring is on its way. The sun is already moving northward and in a few months will bring glorious summer to this corner of England. But before this happens, the tree will start to stir, as it has done a hundred times before. Its roots will absorb moisture and nutrients; channels beneath the bark will pump liquid nourishment to every branch and twig; cells will grow and divide; buds will burst, leaves unfurl. This entire orchestration requires a huge investment by the tree, justified only because it will bring future rewards. During long hot days of sunshine the leafy green canopy—not as luxuriant and vigorous as it once was, but still impressive—will capture energy and use it to drive replenishment and growth.

This tree, like all plants, has an internal clock: its cells perform a set of biochemical reactions that rise and fall every twenty-four hours. The rhythm of these circadian (daily) reactions is normally entrained by the external light but continues even in darkness: flowers open and close on a roughly twenty-four-hour period even in the dark. And by comparing this internal clock with the external periods of light and dark, the tree can judge day length and hence know when to initiate the events of leaf production.

Circadian cycles exist not only in plants but also in cells of all kinds, from bacteria to humans. Biochemical circuits oscillating with periods matching the outside world relieve the organism from the need to continually monitor light levels. They continue to rise and fall even on a cloudy day or if the animal is in a cave.

Nor is it necessary for the environmental variables to show perfectly periodic rhythms. In fact, some fluctuations that seem random when viewed in isolation can, nevertheless, be highly predictable when considered in the context of other parameters. The medley of stimuli encountered by a cell contains significant clues as to what will happen next. For example, a bacterium ingested by a cow or a human usually experiences an immediate increase in temperature. Moreover, this increase in temperature will almost always be followed by a decrease in oxygen levels as the bacterium is carried down into the gastrointestinal tract. The ecology of the mammalian digestive system imposes strong selection for colonization, so if our bacterium is able to anticipate the change, it will have a competitive advantage. And indeed, experiments show that bacteria such as *E. coli* respond to an increase in temperature by shutting down their aerobic metabolism. This built-in reflex works even under laboratory conditions when oxygen is still plentiful. It is a causal link written in biochemistry.

Such coupling of environmental parameters through internal circuits allows a cell to predict future events. These circuits contain, implicitly, the probability of certain life-changing events. They prepare the organism for fluctuations in food and other resources by expressing appropriate actions and mounting protective responses to extreme perturbations. In higher organisms the basic cognitive capacity that underlies predictive behavior is thought to require networks of neurons organized within the architecture of nervous systems. But it also exists in single-cell organisms, implemented by networks of protein interactions.

Each type of response has a characteristic time that matches, or attempts to match, the external constraint. The bacterium cannot know when it will have an opportunity to eat lactose, but when it does, it should be able to marshal its digestive proteins quickly. The crustacean *Artemia,* or brine shrimp, cannot know beforehand when it will experience a period of drought. But when it does, it must act quickly. Arresting its development at a specific stage, the organism produces a resistant spore that can survive for months or years in the complete absence of water. Strategies such as these provide evidence of a long-term memory wherein past experience is written into the genetic score, and such strategies can be surprisingly clever and subtle. Organisms even speculate on the future.

Microbiologists have known since 1944 that when a culture of bacteria is treated with antibiotics such as penicillin, not every cell is killed. The survivors that remain—too numerous to be genetically altered mutants—are, however, only temporarily resistant. Put these "persistent bacteria" back into nutrient medium and allow them to grow again, and they become just as sensitive to the antibiotic as before. Again, most cells are killed by penicillin but a few survive.

Nathalie Balaban, in Stan Leibler's laboratory at Princeton, explored the basis of this phenomenon. Using technical tricks to visualize and follow individual bacteria, she found that cultures of *E. coli* always contain a small number of cells that grow much more slowly than the rest. This is why they are resistant to antibiotics such as penicillin, which target growing cells. But the slow growers are not mutants. They are genetically identical to the normal cells, and the two forms, slow and fast growing, switch from one to the other at low frequency. The detailed molecular mechanism of this switch is still unclear, but its advantages are plain. By coercing a few cells to grow slowly the bacterial culture as a whole takes out an insurance policy. The organism pays a small premium in terms of materials and energy in order to protect itself against future cathartic experiences.

This is not an isolated example, nor is it restricted to bacteria. Wherever investigators look closely, they find evidence of cells deliberately producing variations so as to better meet an uncertain future.

Our amoeba leads such a varied, adventurous life it must surely be at least as clever as a bacterium. It must have sophisticated responses to many environmental challenges, even though we are currently ignorant of what these are. Moreover, the internal representation of the world it carries will include not just static features of the world around it but also temporal relationships. The amoeba will be equipped not only to recognize certain combinations of external stimuli but also to respond optimally when those stimuli are encountered in the correct sequences and intervals.

This is possible because, in the final analysis, biochemical reactions them-selves have an intrinsic tempo and unfold at a characteristic rate. If a cell can match fluctuations in its internal chemistry to events in the outside world, it will be able to anticipate major changes and prepare for them.

So when you observe through a microscope an amoeba crawling on the bottom of a pond, you are watching thousands of protein circuits in action. Receptors on its surface sniff the water for the presence of food, guard for potentially troublesome changes in temperature and acidity, remain alert to vibrations that could indicate the presence of prey, test the substratum for a mechanically suitable foothold. Cascades of cou-pled reactions travel from the receptors into the cell, spread in intersect-ing chemical waves carrying messages and causing changes. Signals arriving at the cytoskeleton modulate the activities of protein circuits re-sponsible for movement, cytoplasmic flow, and extension or retraction.

All of these reactions are programmed into the genetic instructions of DNA. But in another sense they are highly individual and unpre-dictable. Every living cell will have its own chemical fingerprint, shaped by its particular circumstances. The cascade of reactions in the leading region of a hungry amoeba will differ from those in the trailing tail of one that is fully fed. Within each amoeba will be proteins that would—if you were able to read them correctly—tell you about its present state and recent past. One protein might control the paths of metabolism, ad-justing them to the recent diet. Others might change with the life stage of the amoeba. A greedy juvenile whose only concern is to feed as fast as possible will not be the same as a mature cell about to divide.

In her 1983 Nobel Prize acceptance speech, the American geneticist Bar-bara McClintock identified a goal for future biologists: "To determine the extent of knowledge the cell has of itself." We are still a long way from achieving this goal, but it is at least now clear in what form this knowledge is stored. Protein molecules represent, in all their environment-tasting, message-generating, feature-extracting, memory-storing prodigality, what a cell knows of itself.

But there was a second part to Barbara McClintock's quest. She also wanted to know how a cell uses its knowledge of the world in a thoughtful manner when challenged. A thousand chemical voices clamor for attention every instant of an amoeba's existence. How does it decide what to do? The question seems especially urgent for a migrating cell, since it must instantly choose to go this way rather than that. Should it move toward a potential source of food even if this also carries it into an unpleasant acidic neighborhood? Is it better to follow a potentially edible spore that has eluded attempts at capture over the past five minutes, or to strike out in a new direction in the hope that this will bring better returns?

Other decisions will also have to be made that have nothing to do with motility. Should the cell make this or that set of proteins? Which of several alternative pathways of metabolism should it adopt? And related to these questions is the altogether more philosophical issue of what exactly I mean by saying that "a cell makes a decision." Doesn't this call for a central executive, some body to which all information ultimately flows and from which all directives emanate? Natural selection of ancestral amoebae must have installed a basis for such judgments. How and where are these priorities expressed in physical terms?

It may be helpful to briefly consider how humans process incoming stimuli. Richard Gregory, an eminent expert on the psychology of human vision, distinguishes between the reception of stimuli by the brain and its perception. Developing earlier notions of the nineteenth-century German physician and physicist Hermann von Helmholtz, Gregory defines reception as the raw sensory stimuli that are received in unprocessed format and perception as the effect of such input after processing in the brain. Vision, for example, is not due to an actual image projected onto the brain as if on a screen. The visual scene is actually fragmented into different features and then put back together again. Light enters the retina, where it activates up to 120 million photosensitive cells (rods and cones). These are reduced to about 1 million axons in the optic nerve before being carried to the primary visual cortex. There the signals are broken down into separate features—orientation, color, movement—before eventually being recombined. The internal "view" thus created is based on a hypothesis of what the outside world is like and can be easily duped—for example, by optical illusions. Note that light itself never enters the brain. No matter how dramatic the visual experience, it is completely dark in there—just pulses of electricity moving along axons.

Something like this happens in a single cell. The raw stimuli it experiences are also fragmented, albeit into chemical rather than electrical signals (a minor distinction). I discussed in Chapter 6 the semantic content of proteins: how they "mean" certain things to the cell. What I did not consider was what would happen if these semantic proteins themselves talked to each other. What if a protein representing hunger, a protein representing detection of food, and a protein that tells the cell that it is attached to a suitable matrix for locomotion all send signals to the same integrating protein? Would not this common target encapsulate into the way it handles this confluence of signals a description of the outside world?

Consider a "skydiving" amoeba (Chapter 1) that has become detached from the surface and now adopts a spread-eagle form as it searches for a new surface. Imagine a loose consortium of protein complexes, one for each pseudopodium, each a transient, temporary affair, newly created and easily dispersed once the pseudopodium has retracted. Each pseudopodium *could* act as an independent agent that tests its local environment. Each could sniff for food and other chemical signals and in consequence develop weaker or stronger adhesions—pull more or less strongly against its adhesions. The cell as a whole could then be led by the strongest pseudopodium, a simple matter of mechanics without need for higher control. The appearance of motivation and intent would then arise, as in the Genghis robot, from the collective action of a set of tiny computers without overarching control. But considering all the other remarkable things a cell can do, I wonder whether a more sophisticated mechanism might not have evolved.

Would it not be an efficient solution if the molecular circuits activated as an amoeba settles onto a surface in some way represent the shape of the cell and contain some notion of the features it was seeking? Surely it is possible that local controllers in different parts of the cell send signals to a single executive complex. This pontifical center, the equivalent in protein terms of a grandmother nerve cell, would receive relevant signals about the cell's internal state (metabolic level, position in the cell cycle, level of activity of organelles), as well as signals filtering from the outside (via membrane receptors relaying mechanical and chemical stimuli). Logical processes of the kind described previously could then combine these signals and compare their strengths and timings. The results of these computations—these executive decisions—would fan out to different regions of the cell and control their several movements.

And where in the cell might such a center be located? The logical place in an animal cell would be its centrosome. This collection of protein and RNA molecules typically sits beside the nucleus—close to the center of the cell, as its name implies. The centrosome acts as a seed for microtubules that grow out to other parts of the cell, including its membrane, and thereby influence both the shape of the cell and its movements. It plays a crucial role in cell division and the segregation of chromosomes to the daughter cells. Enigmatic structures called centrioles embedded in the centrosome direct their own replication and have been observed wandering away from the centrosome to distant locations near the cell membrane. There is no evidence so far as I know that the centrosome contributes to other cell functions such as metabolism or biosynthesis, but it could.

You might think that this is already carrying speculation too far. But as André Gide wrote: "One doesn't discover new lands without consenting to lose sight of the shore for a very long time." So let me, in this spirit, take one further step and ask what implications all this talk of bacteria and amoebae has for human mentation? What, if anything, does an understanding of the computational processes inside a single cell say about our knowledge, our cognition, and our awareness?

Reflecting on this unpromising question, I realize that I have so far assumed that I know what these terms mean for humans and then employed them as a basis to assess processes in cells. But is this justified? How well could I define, if put to the task, a visual percept of an orange or an orangutan in a human brain? Not very well—certainly not at the level of particular nerve cells or synapses, or patterns of electrical signals. But what if I turned the flashlight back on itself and used cell biology to illuminate psychology? Many processes such as respiration and movement have common features in both humans and micro-organisms. They have the same underlying molecular basis and can be studied in whatever organism is most convenient, which often turns out to be the smallest and simplest.

So you could argue that we understand the recognition of features of the external world, the storage of memories, the influence of chance events on behavior, the prediction of future events, better for a cell than we do for a higher animal. Given the long-standing correspondence be-tween the two because of their common evolutionary origins, it might be, as Dale Purves, professor of neuroscience at Duke University, said to me, that cells are "touchstones for human mentation." A touchstone is a slab of dark mineral used to test the quality of gold and silver from the color of the streak each produces on it . . . not a bad analogy.

Glossary

ACTIN: Abundant protein that forms thin filaments in eukaryotic cells and plays a major role in their movements.

ACTION POTENTIAL: Rapid, transient, self-propagating electrical signal in the membrane of a nerve cell.

ADAPTATION: Decrease in sensitivity to a repeated or prolonged stimulus.

ALGA (plural *algae*): One of a wide range of photosynthetic organisms, such as Chlamydomonas and Volvox.

ALGORITHM: Explicit instructions for the performance of a logical task or solution of a mathematical problem.

ALLOSTERY: Ability of a protein molecule to switch from one shape to another depending on the binding of a regulatory molecule. From Greek *allos,* other, and *stereos,* shape.

AMINO ACIDS: Small molecules containing both an amino group and a carboxyl group that can be strung together to make proteins.

AMOEBA (plural *amoebae*): Free-living single-celled eukaryotic organism that crawls by changing its shape.

AMOEBA PROTEUS: Species of giant freshwater amoeba often used in studies of cell locomotion.

ANTIBODY: Protein produced by white blood cells in response to an infection. Attaches tightly to the foreign molecule or cell, thereby inactivating it or marking it for destruction.

ANTIGEN: Molecule that provokes the cellular production of specific antibodies in an immune response.

ARTIFICIAL INTELLIGENCE: Computer software developed to mimic such human intelligent behavior as playing chess or recognizing shapes and faces.

ASPARTATE: One of the twenty naturally occurring amino acids: a component of proteins.

ATP (adenosine 5′-triphosphate): Small molecule that is the principal carrier of chemical energy in cells.

AXON: Long nerve cell process that is capable of rapidly conducting nerve impulses. See *action potential*.

BASE PAIRING: Set of hydrogen bonds that hold together specific pairs of nucleotides in an RNA or DNA molecule: G pairs with C, and A with T or U.

BIOFILM: Complex aggregate of bacteria held together by a sticky matrix of polysaccharides and other molecules.

BOOLEAN: System of logic or algebra that defines the combination of binary selections or switches, using operators such as AND, OR, and NOT.

CALMODULIN: Protein that binds calcium ions.

CAM II KINASE (calmodulin-activated protein kinase): Protein in a synapse that responds to changes in calcium ions caused by electrical stimulation.

CATALYST: Substance that speeds a chemical reaction without itself being changed.

CELL BODY: Main part of a nerve cell, containing the nucleus. The other parts are axons and dendrites.

CENTROSOME: Centrally located organelle of an animal cell that is the primary microtubule-organizing center and acts as the spindle pole during mitosis.

CHEMOTAXIS: Movement of a cell or organism toward or away from a diffusible chemical.

CILIUM (plural *cilia*): Hairlike extension on the surface of a cell capable of performing repeated beating movements.

CONFORMATION: Spatial location of the atoms of a molecule; the precise shape of a protein or RNA molecule in three dimensions.

CONFORMATIONAL CHANGE: Switchlike transition in the shape of a protein molecule, often triggered by an external stimulus, such as the binding of an ion or small molecule.

CONSCIOUSNESS: Part of the human mind that is aware of a person's self and responsible for free will. Creates a conceptual private space.

CREATINE KINASE: Enzyme involved in energy metabolism.

CYCLIC AMP: Small molecule generated from ATP widely used as a signaling molecule.

CYTOCHROME C: Protein that transfers electrons during cellular respiration and photosynthesis.

CYTOPLASM: Contents of a cell within its plasma membrane but outside the nucleus (if it has one).

CYTOSKELETON: Extensive system of protein filaments that enables a eukaryotic cell to organize its interior and to perform directed movements.

DENDRITE: Extension of a nerve cell, typically branched and relatively short, that receives stimuli from other nerve cells.

DENDRITIC SPINE: Small projection on the surface of a nerve cell dendrite that receives signals from an incoming axon: the receiving half of a synapse.

DICTYOSTELIUM (dictyostelium discoideum): Cellular slime mold widely used in the study of cell locomotion, chemotaxis, and differentiation.

DIFFERENTIATION: Process by which a cell undergoes a progressive change to a more specialized and usually easily recognized cell type.

DIFFUSION: Net drift of molecules toward regions of lower concentrations, due to random thermal movement.

DNA (DEOXYRIBONUCLEIC ACID): Self-replicating molecule that is the principal carrier of hereditary information in cells.

ENZYME: Biological catalyst; almost always a protein molecule.

EPITHELIUM: Sheet of one or more layers of cells covering an external surface or lining a cavity.

ESCHERICHIA COLI (*E. coli*): Common bacterium of the colon of humans and other mammals, widely used in biomedical research.

EUKARYOTE: Organism composed of one or more cells with a distinct nucleus and cytoplasm. Includes all forms of life except viruses and bacteria.

FEEDBACK INHIBITION: Type of regulation of metabolism in which an enzyme acting early in a reaction pathway is inhibited by a late product of that pathway.

FLAGELLUM (plural *flagella*): Long, whiplike protrusion whose undulations drive a cell through a fluid medium.

FUZZY LOGIC: Reasoning that is approximate rather than precise, typically expressed in such terms as *slightly, quite,* and *very.*

GENETIC ALGORITHM: A search technique used by computer programmers that was inspired by evolutionary biology.

GLUCAGON: Hormone that stimulates release of the sugar glucose into the blood.

GLUTAMATE: One of the amino acids found in proteins. Also used to carry signals in many synapses. See *neurotransmitter.*

HISTONE: Small, positively charged protein associated with DNA in eukaryotic chromosomes.

HISTONE CODE: Pattern of acetyl, phosphate, or other groups added to histone molecules to flag positions of important features of a chromosome.

HYBRIDIZE: Process by which single strands of RNA or DNA find and selectively pair with other strands having a complementary sequence of bases.

IMMUNOGLOBULIN: Protein with antibody activity produced by a lymphocyte as part of the immune defense.

KINASE: Enzyme that adds a phosphate group to another protein.

LIPID BILAYER: Thin sheet of lipid that forms the structural basis for all cell membranes.

LUCIFERASE: Enzyme responsible for light production in luminescent organisms such as fireflies.

LYMPHOCYTE: White blood cell that makes an immune response when activated by a foreign molecule (an antigen).

MACROMOLECULE: Very large molecule such as a protein, nucleic acid, or polysaccharide made as a polymer of smaller chemical units.

MACROPHAGE: White blood cell that is specialized for the uptake and digestion of particles such as dead cells or invading organisms.

MEMBRANE: Thin sheet of lipid molecules and associated proteins that encloses all cells and forms the boundaries of many eukaryotic organelles.

MEMORY: Change in state of a cell or organism owing to external events that is used in a meaningful way to determine subsequent actions.

METHYL: Chemical group with the formula CH_3.

METHYLATION: Addition of a methyl group, for example to a protein.

MICRON OR MICROMETER (μm): Unit of length used to measure cells. One micron equals one thousandth of a millimeter; eighty microns equal the width of a typical human hair.

MICROORGANISM: Organism of microscopic size such as a bacterium or protozoan.

MICROTUBULES: Long, stiff protein filaments that contribute to cell shape and provide tracks for intercellular movement.

MITOCHONDRION (plural *mitochondria*): Membrane-bounded organelle, about the size of a bacterium, that performs the last stages of breakdown of food and produces most of the ATP in eukaryotic cells.

MORPHOGENESIS: Development of form and structure in a cell or organism, usually in reference to embryonic development.

MOTOR PROTEIN: Protein that uses energy derived from ATP to propel itself along a protein filament (an actin filament or microtubule).

MULTICELLULAR ORGANISM: Living creature made of many cells, including most plants and animals.

MYOSIN: Motor protein that drives movements along actin filaments.

MYXOBACTERIA: Type of bacteria that can aggregate into multicellular fruiting bodies containing resting spores.

NANOMETER (nm): Unit of length used to measure molecules. One nanometer equals one billionth of a meter, one thousandth of a micron.

NEURAL NETWORK: Computer-based web of processing elements analogous to real nerve cells used in various tasks of pattern recognition.

NEUREXIN: Protein involved in cell adhesion in the nervous system.

NEURON OR NERVE CELL: Cell with long processes specialized to receive, conduct, and transmit signals in the nervous system.

NEUROTRANSMITTER: Small molecule such as glutamate that is secreted by a nerve cell at a synapse to signal to the adjoining cell.

NOISE: Unexpected, random fluctuations in a chemical or physical process due to thermal energy.

ORGANELLE: Functional unit of a cell, like a small organ.

PARAMECIUM (plural *paramecia*): Freshwater protozoan having an oval body covered with cilia.

PEPTIDE: Molecule made from two or more amino acids linked head to tail.

PERCEPTRON: Computer device that can be trained to produce a specified output from a given set of inputs.

PHAGOCYTOSIS: Process by which particulate material is engulfed ("eaten") by a cell.

PHOSPHATASE: Enzyme that removes a phosphate group from a protein (opposite of a kinase).

PHOSPHATE: Common chemical compound containing phosphorus and oxygen, formula PO_4.

PHOSPHORYLATE: Addition of a phosphate group to a molecule, usually from ATP by means of an enzyme. See *kinase*.

PLASMA MEMBRANE: Membrane that surrounds a living cell.

POLYMER: Large molecule made by linking many smaller units (monomers).

POSTSYNAPTIC: Pertaining to the part of a synapse that receives the transmitted signal, as in the postsynaptic cell and postsynaptic density. See also *dendritic spine*.

PRESYNAPTIC: Pertaining to the part of a synapse that sends a signal.

PROKARYOTE: Single-celled microorganism whose simple cells lack a well-defined, membrane-enclosed nucleus. Mainly bacteria.

PROTEIN: One of many large molecules made as a chain of amino acids, enormously variable in size and shape and with a vast range of chemical and physical properties.

PROTEIN COMPLEX: Group of protein molecules held together by weak chemical bonds and performing a specific task.

PROTOZOAN (plural *protozoa*): Free-living, single-celled, motile eukaryotic organism, such as paramecia or amoebae.

PSEUDOPODIUM (plural *pseudopodia*): Large surface protrusion formed by amoeboid cells as they crawl.

QUORUM SENSING: Ability of bacteria to sense and influence the population to which they belong through the secretion of small molecules.

RECEPTOR: Protein that detects an environmental stimulus such as a change in concentration of a specific substance and then initiates a response in the cell.

RIBOSOME: Particle made of RNA and proteins that "reads" the sequence of bases in messenger RNA and translates them into protein.

RIBOZYME: RNA molecule able to catalyze a chemical reaction.

RNA (ribonucleic acid): Linear polymer formed from chains of nucleotide bases. Made as a copy of a limited region of DNA.

RNA POLYMERASE: Enzyme that moves along a DNA molecule copying the sequence into an RNA molecule.

ROBOT: Machine in human form or with humanlike features, and possessing a degree of intelligence. From Czech *robota*, compulsory worker.

SECRETION: Process by which molecules are released from a eukaryotic cell.

SIGNALING MOLECULE: Molecule that changes its concentration in response to an environmental stimulus.

SIGNAL TRANSDUCTION: The process by which a cell detects a physical or chemical stimulus and converts it into a molecular change.

SLIME MOLD: Primitive organism that forms spores, like fungi, but feeds on bacteria, like protozoa.

STENTOR: Freshwater protozoan with a trumpet-shaped body covered with cilia.

SUBSTRATE: Molecule on which an enzyme acts.

SYNAPSE: Specialized junction between nerve cells across which the nerve impulse is transferred. In most synapses the signal is carried by a neurotransmitter, which is secreted by the input cell and diffuses to the target.

SYNAPTIC VESICLE: Small membrane-bounded organelle found in a synapse that contains neurotransmitter.

TEMPLATE: Single strand of DNA or RNA whose sequence of bases acts as a guide for the synthesis of a complementary strand.

VESICLE: Small, membrane-bounded, spherical organelle in the cytoplasm of a eukaryotic cell. See *synaptic vesicle*.

VITALISM: Belief that living processes can be explained only by something outside physics and chemistry, namely, a vital force.

WETWARE: Collective term for the information-rich, computational processes found in living cells.

Sources and Further Reading

ONE

Clever Cells

Alberts, B., D. Bray, K. Hopkins, A. Johnson, J. Lewis, M. Raff, K. Roberts, and P. Walter. 2004. *Essential Cell Biology*. 2nd ed. New York: Garland Science.

Berg, H. C. 2004. *E. coli in Motion*. New York: Springer.

Buchsbaum, R. 1976. *Animals Without Backbones*. Chicago: University of Chicago Press.

Gibbs, D. 1908. "The Daily Life of Amoeba proteus." *American Journal of Psychology* 19:232–41.

Grimstone, A. V., and L. R. Cleveland. 1965. "The Fine Structure and Function of the Contractile Axostyles of Certain Flagellates." *Journal of Cell Biology* 24:387–400.

Hodge, C. F., and H. A. Aikens. 1895. "The Daily Life of a Protozoan: A Study in Comparative Psycho-Physiology." *American Journal of Psychology* 6: 524–33.

Jennings, H. S. 1906. *The Behavior of the Lower Organisms*. New York: Columbia University Press.

Loeb, J. 1918. *Forced Movements, Tropisms and Animal Conduct*. Philadelphia: Lippincott.

Morimoto, B. H., and D. E. Koshland. 1991. "Short-Term and Long-Term Memory in Single Cells." *Journal of the Federation of American Societies for Experimental Biology* 5:2061–67.

Nakagaki, T., H. Yamada, and M. Hara. 2004. "Smart Network Solutions in an Amoeboid Organism." *Biophysical Chemistry* 107:1–5.

Pfeffer, W. 1888. "Über chemotaktische Bewegungen von Bacterien, Flagellaten, und Volvocineen." *Untersuchungen aus dem Botanischen Institut in Tübingen*. 2:584–88.

Saville-Kent, W. 1880. *A Manual of the Infusoria*. Vol. 1. London: David Bogue.

Tartar, V. 1961. *The Biology of Stentor*. Oxford: Pergamon.

Woolridge, D. E. 1963. *The Machinery of the Brain*. New York: McGraw-Hill.

TWO
Simulated Life

Bartlett, F. C. 1932. *Remembering: A Study in Experimental and Social Psychology.* Cambridge: Cambridge University Press.

Carter, R. 2002. *Exploring Consciousness.* Los Angeles: University of California Press.

Castronova, E. 2005. *Synthetic Worlds: The Business and Culture of Online Games.* Chicago: University of Chicago Press.

Feynman, R. P. 1999. *Feynman Lectures on Computation.* New York: Basic.

Hofstadter, D. R., and D. C. Dennett. 1981. *The Mind's I: Fantasies and Reflections on Self and Soul.* New York: Basic.

Miedaner, T. 1977. *The Soul of Anna Klane.* Pearce, Ariz.: Church of Physical Theology.

Ramachandran, V. S. 1998. "Consciousness and Body Image: Lessons from Phantom Limbs, Capgras Syndrome and Pain Asymbolia." *Philosophical Transactions of the Royal Society of London* B 353:1852–59.

Ramachandran, V. S., and S. Blakeslee. 1998. *Phantoms in the Brain.* London: Fourth Estate.

Sacks, O. 1985. *The Man Who Mistook His Wife for a Hat.* London: Gerald Duckworth.

Walter, W. G. 1950. "An Imitation of Life." *Scientific American* 182:42–45.

———. 1951. "A Machine That Learns." *Scientific American* 185:60–63.

———. 1953. *The Living Brain.* New York: W. W. Norton.

Whitesides, G. M. 2003. "The 'Right' Size in Nanobiotechnology." *Nature Biotechnology* 21:1161–65.

THREE
Protein Switches

Berg, H. C. 1993. *Random Walks in Biology.* Princeton, N.J.: Princeton University Press.

Bray, D. 1995. "Protein Molecules as Computational Elements in Living Cells." *Nature* 376:307–12.

Brown, G. 1999. *The Energy of Life.* London: HarperCollins.

Edsall, J. T. 1987. "Hemoglobin and the Origins of the Concept of Allosterism." *Federation Proceedings* 39:227–35.

Fersht, A. 1985. *Enzyme Structure and Mechanism.* 2nd ed. New York: W. H. Freeman.

Judson, H. F. 1979. *The Eighth Day of Creation.* London: Jonathan Cape.

Monod, J., and F. Jacob. 1961. "General Conclusions: Telenomic Mechanisms in Cellular Metabolism, Growth, and Differentiation." *Cold Spring Harbor Symposia on Quantitative Biology* 26:389–401.

Perutz, M. F. 1989. "Mechanisms of Cooperativity and Allosteric Regulation in Proteins." *Quarterly Reviews of Biophysics* 22:139–237.

Shacter, E., P. B. Chock, and E. R. Stadtman. 1984. "Regulation Through Phosphorylation/Dephosphorylation Cascade Systems." *Journal of Biological Chemistry* 259:12252–59.

Umbarger, H. E. 1992. "The Origin of a Useful Concept: Feedback Inhibition." *Protein Science* 1:1392–95.

FOUR

Protein Signals

Alon, U. 2007. *An Introduction to Systems Biology.* Boca Raton, Fla.: Chapman and Hall/CRC.

Arkin, A., and J. Ross. 1994. "Computational Functions in Biochemical Reaction Networks." *Biophysical Journal* 67:560–78.

Brandman, O., and T. Meyer (2008). "Feedback Loops Shape Cellular Signals in Space and Time." *Science* 322:390–95.

Bray, D. 2001. *Cell Movements: From Molecules to Motility.* 2nd ed. New York: Garland.

Mangan, S., and U. Alon. 2003. "Structure and Function of the Feed-Forward Loop Network Motif." *Proceedings of the National Academy of Sciences of the United States of America* 100:11980–85.

McCulloch, W. S., and W. Pitts. 1943. "A Logical Calculus of the Ideas Immanent in Nervous Activity." *Bulletin of Mathematical Biophysics* 5:115–33.

Mettetal, J. T., D. Muzzey, C. Gómez-Uribe, and A. van Oudenaarden. 2008. "The Frequency Dependence of Osmo-Adaptation in *Saccharomyces cerevisiae.*" *Science* 319:482–84.

Roach, P. J. 1990. "Control of Glycogen Synthase by Hierarchical Protein Phosphorylation." *Journal of the Federation of American Societies for Experimental Biology* 4:2961–68.

Rust, M. J., J. S. Markson, W. S. Lane, D. S. Fisher, and E. K. O'Shea. 2007. "Ordered Phosphorylation Governs Oscillation of a Three-protein Circadian Clock." *Science* 318:809–12.

Wilson, W. A., et al. 2005. "Control of Mammalian Glycogen Synthase." *Proceedings of the National Academy of Sciences of the United States of America* 102:16596–16601.

FIVE

Cell Wiring

Alberts, B. 1998. "The Cell as a Collection of Protein Machines." *Cell* 92:291–94.

Berg, H. C. 1988. "A Physicist Looks at Bacterial Chemotaxis." *Cold Spring Harbor Symposia on Quantitative Biology* 53:1–9.

———. 2004. *E. coli in Motion.* New York: Springer.

Church, J. 1961. *Language and the Discovery of Reality.* New York: Random House.

Darwin, C. 1859. *The Origin of Species.* London: Penguin.

Gerhart, J., and M. Kirschner. 1997. *Cells, Embryos, and Evolution.* Oxford: Blackwell Science.

Hanczyc, M. M., S. M. Fujikawa, and J. W. Szostak. 2003. "Experimental Models of Primitive Cellular Compartments: Encapsulation, Growth, and Division." *Science* 302:618–22.

Jablonka, E., and M. J. Lamb. 1995. *Epigenetic Inheritance and Evolution.* Oxford: Oxford University Press.

Mandelbrot, B. B. 1982. *The Fractal Geometry of Nature.* San Francisco: W. H. Freeman.

Raser, J. M., and E. K. O'Shea. 2005. "Noise in Gene Expression: Origins, Consequences, and Control." *Science* 309:2010–13.

Rosenthal, J. J., and F. Bezanilla. 2002. "Extensive Editing of mRNAs for the Squid Delayed Rectifier K+ Channel Regulates Subunit Tetramerization." *Neuron* 34:743–57.

Salman, H., and A. Libchaber. 2007. "A Concentration-Dependent Switch in the Bacterial Response to Temperature." *Nature Cell Biology* 9:1098–1100.

Thomas, L. 1979. *The Medusa and the Snail: More Notes of a Biology Watcher.* New York: Viking Press.

SIX

Neural Nets

Baas, A. F., et al. 2004. "Complete Polarization of Single Intestinal Epithelial Cells Upon Activation of LKB1 by STRAD." *Cell* 116:457–66.

Barabási, A.-L. 2002. *Linked: The New Science of Networks.* Cambridge, Mass.: Perseus.

Bray, D., and S. Lay. 1994. "Computer Simulated Evolution of a Network of Cell-Signaling Molecules." *Biophysical Journal* 66:972–77.

Carling, D. 2004. "The AMP-Activated Protein Kinase Cascade: A Unifying System for Energy Control." *Trends in Biochemical Science* 29:18–24.

Crick, F. H. C. 1994. *The Astonishing Hypothesis: The Scientific Search for the Soul.* New York: Touchstone.

Jost, C. R., et al. 2002. "Creatine Kinase B-Driven Energy Transfer in the Brain Is Important for Habituation and Spatial Learning Behaviour, Mossy Fibre Field Size, and Determination of Seizure Susceptibility." *European Journal of Neuroscience* 15:1692–1706.

Noble, D. 2006. *The Music of Life: Biology Beyond the Genome.* Oxford: Oxford University Press.

Sejnowski, T. J., and C. R. Rosenberg. 1987. "Parallel Networks That Learn to Pronounce English Text." *Complex Systems* 1:145–68.

SEVEN
Cell Awareness

Brown, G. 1999. *The Energy of Life.* London: HarperCollins.
Gregory, R. L. 1998. *Eye and Brain: The Psychology of Seeing.* 5th ed. Oxford: Oxford University Press.
Harold, F. M. 1995. *The Vital Force: A Study of Bioenergetics.* New York: W. H. Freeman.
Hille, B. 1992. *Ionic Channels of Excitable Membranes.* 2nd ed. Sunderland, Mass.: Sinauer.
Holmes, F. L. 1985. *Lavoisier and the Chemistry of Life.* Madison: University of Wisconsin Press.
Jaynes, J. 1976. *The Origin of Consciousness in the Breakdown of the Bicameral Mind.* Boston: Houghton Mifflin.
Wundt, W. M., and E. B. Titchener. 1910. *Principles of Physiological Psychology.* New York: Macmillan.

EIGHT
Molecular Morphing

Gilbert, W. 1986. "The RNA World." *Nature* 319:618.
Kirschner, M. W., and J. Gerhart. 2005. *The Plausibility of Life: Resolving Darwin's Dilemma.* New Haven: Yale University Press.
Miller, S. L., and H. Urey. 1959. "Organic Compound Synthesis on the Primitive Earth." *Science* 130:245–51.
Orgel, L. E. 1994. "The Origin of Life on Earth." *Scientific American* 271: 76–83.
Parham, P. 2000. *The Immune System.* New York: Garland.
Ridley, M. 1994. *The Red Queen.* New York: Macmillan.
Watson, J. D. 1993. "Prologue: Early Speculations and Facts About RNA Templates." In *The RNA World*, ed. R. F. Gesteland and J. F. Atkins, xv–xxiii. Cold Spring Harbor, N.Y.: Cold Spring Harbor Laboratory Press.

NINE
Cells Together

Bassler, B. L., and R. Losick. 2006. "Bacterially Speaking." *Cell* 125:237–46.
Coburn, P. S., et al. 2004. "Enterococcus faecalis Senses Target Cells and in Response Expresses Cytolysin." *Science* 306:2270–72.
Lenhoff, H. M., and S. G. Lenhoff. 1988. "Trembley's Polyps." *Scientific American* 258:86–91.
Tyson, J. J., K. C. Chen, and B. Novak. 2003. "Sniffers, Buzzers, Toggles, and Blinkers: Dynamics of Regulatory and Signaling Pathways in the Cell." *Current Opinion in Cell Biology* 15:221–31.

TEN
Genetic Circuits

Dekel, E., and U. Alon. 2005. "Optimality and Evolutionary Tuning of the Expression Level of a Protein." *Nature* 436:588–92.

Elowitz, M. B., and S. Leibler. 2000. "A Synthetic Oscillatory Network of Transcriptional Regulators." *Nature* 403:335–38.

Gardner, T. S., C. H. Cantor, and J. J. Collins. 2000. "Construction of a Genetic Toggle Switch in Escherichia coli." *Nature* 403:339–42.

Istrail, S., S. B.-T. de Leon, and E. H. Davidson. 2007. "The Regulatory Genome and the Computer." *Developmental Biology* 310:187–95.

Jacob, F. 1987. *La Statue Intérieure*. Paris: Gallimard.

Kauffman, S. 1974. "The Large Scale Structure and Dynamics of Gene Control Circuits: An Ensemble Approach." *Journal of Theoretical Biology* 44:167–90.

———. 1995. *At Home in the Universe: The Search for the Laws of Self-Organization and Complexity*. New York: Oxford University Press.

Lee, T. I., et al. 2002. "Transcriptional Regulatory Networks in Saccharomyces cerevisiae." *Science* 298:799–804.

Levine, M., and E. H. Davidson. 2005. "Gene Regulatory Networks for Development." *Proceedings of the National Academy of Sciences of the United States of America* 102:4936–42.

Xiong, W., and J. E. Ferrell. 2003. "A Positive-Feedback-Based Bistable 'Memory Module' That Governs a Cell Fate Decision." *Nature* 426:460–64.

ELEVEN
Robots

Bonabeau, E., M. Dorigo, and G. Theraulaz. 1999. *Swarm Intelligence: From Natural to Artificial Systems*. New York: Oxford University Press.

Brooks, R. A. 2002. *Robot: The Future of Flesh and Machines*. London: Penguin.

Grand, S. 2000. *Creation: Life and How to Make It*. London: Weidenfeld and Nicolson.

———. 2003. *Growing Up with Lucy: How to Build an Android in Twenty Easy Stages*. London: Weidenfeld and Nicolson.

Ijspeert, A. J., A. Crespi, D. Ryczko, and J.-M. Cabelguen. 2007. "From Swimming to Walking with a Salamander Robot Driven by a Spinal Cord Model." *Science* 315:1416–20.

Lambrinos, D., et al. 2000. "A Mobile Robot Employing Insect Strategies for Navigation." *Robotics and Autonomous Systems* 30:39–64.

Theraulaz, G., J. Gautrais, S. Camazine, and J.-L. Deneubourg. 2003. "The Formation of Spatial Patterns in Social Insects: From Simple Behaviors to Complex Structures." *Philosophical Transactions of the Royal Society of London* A 361:1263–82.

Webb, B. 2002. "Robots in Invertebrate Neuroscience." *Nature* 417:359–63.

TWELVE

The Juice

Adrian, E. D. 1934. *The Basis of Sensation.* London: Christophers.

Bhalla, U. S., and R. Iyengar. 1999. "Emergent Properties of Networks of Biological Signaling Pathways." *Science* 283:381–87.

Changeux, J.-P. 1983. *L'homme neuronal.* Paris: Hachette Littérature.

Hebb, D. O. 1949. *The Organization of Behavior.* New York: Wiley.

Kennedy, M. B. 2000. "Signal-Processing Machines at the Postsynaptic Density." *Science* 290:750–54.

Landauer, T. K. 1986. "How Much Do People Remember? Some Estimates of the Quantity of Learned Information in Long-Term Memory." *Cognitive Science* 10:477–93.

Lisman, J. E., H. Schulman, and H. Cline. 2002. "The Molecular Basis of CaMKII Function in Synaptic and Behavioural Memory." *Nature Reviews Neuroscience* 3:175–90.

Maynard Smith, J. 2000. "The Concept of Information in Biology." *Philosophy of Science* 67:177–94.

McCorduck, P. 2004. *Machines Who Think.* 2nd ed. Natick, Mass.: A. K. Peters.

Miller, P., A. M. Zhabotinsky, J. E. Lisman, and X.-J. Wang. 2005. "The Stability of a Stochastic CaMKII Switch: Dependence on the Number of Enzyme Molecules and Protein Turnover." *PLoS Biology* 3:705–17.

Spruston, N. 2008. "Pyramidal Neurons: Dendritic Structure and Synaptic Integration." *Nature Reviews Neuroscience* 9:206–21.

von Neumann, J. 1958. *The Computer and the Brain.* New Haven: Yale University Press.

THIRTEEN

Amoeba Redux

Albrecht-Buehler, G. 2008. "Cell Intelligence." Available at www.basic.northwestern.edu/g-buehler/cellinto.htm.

Balaban, N. Q., J. Merrin, et al. 2004. "Bacterial Persistence as a Phenotypic Switch." *Science* 305:1622–25.

Gregory, R. L. 1997. "Knowledge in Perception and Illusion." *Philosophical Transactions of the Royal Society of London* B 352:1121–28.

Losick, R., and C. Desplan. 2008. "Stochasticity and Cell Fate." *Science* 320: 65–68.

Purves, D., and R. B. Lotto. 2003. *Why We See What We Do: An Empirical Theory of Vision.* Sunderland, Mass.: Sinauer Associates.

Ramanathan, S., and J. R. Broach. 2007. "Do Cells Think?" *Cellular and Molecular Life Sciences*. 64:1801–04.

Samoilov, M. S., G. Price, and A. P. Arkin. 2006. "From Fluctuations to Phenotypes: The Physiology of Noise." *Science STKE* 366 (December 19):re17.

Singer, S. J., and G. L. Nicolson. 1972. "The Fluid Mosaic Model of the Structure of Cell Membranes." *Science* 175:720–31.

Tagkopoulos, I., Y.-C. Liu, and S. Tavazoie. 2008. "Predictive Behavior Within Microbial Genetic Networks." *Science* 320:1313–17.

Index

Church, Joseph, 100–101
ciliates, 14–18
circadian rhythm, 234–35
Cold Spring Harbor Laboratories meeting
  (1961), 60–61, 63
Collins, James, 181–82
combinatorial units, 116
communication. *See* cell communication
complexity: amoeba and, 236–37; of genetic circuits, 184–86; hierarchical organization and, 230–33; of individual cells,
  229–30; Kauffman on, 189–90; of nerve
  cells, 222; of protein molecules, 229–30
computation: electronic vs. protein molecules, 87; living cells and, 61; by nervous systems, 194; protein circuits and,
  75–77, 79–82; sensory evaluation, 72; in
  single-celled organisms, 227; synapses
  and, 222; theory of, 39–40
computational units (neural networks),
  112–16, *114*
The Computer and the Brain (von Neumann), 211–12
computer games: avatars, 44–45; FPS
  (first-person shooting games), 44–45;
  Norns, 197–99; PacMan, 28–31, *30,* 39,
  40; virtual reality, 46–47
computer models, 99–104
computer programs: comparison to gene
  circuits, 184; incorporation of personal
  philosophy in, 198–99; as language,
  100–101; NETtalk, 109–10, 112
computers: comparisons to brains, 27–28,
  210–11; digital/analog technology,
  82–84; living cells as, 40–41, 227–28;
  sentience of, ix–x; simulation of bacterial chemotaxis, 99–104; use of term,
  40–41; virtual reality, 46–47. *See also*
  computer games; neural networks; robots
consciousness, 20–26, 51, 132–37, 143. *See
  also* awareness
contact guidance, 171
creatine kinase, 128–30
Creation: Life and How to Make It
  (Grand), 198–99
Crick, Francis, 145, 147
CyberDog, 200

cyclic AMP, 118, 185
cyclic reactions, 75–77, *76,* 90
cytochrome c, 141–42
cytoplasm, 5, 11, 55–56, 73, 92, 104–5
cytosine (C), 146–50, *148*
cytoskeleton, 86

"The Daily Life of Amoeba Proteus"
  (Gibbs), 13–14
Darwin Lecture Series (1996), 109–10
Davidson, Eric, 191
dendrites, 209–10, 213, 214, *214*
dendritic spines, 214, *214*
dephlogistated air, 137
Descartes, René, 20–21
Dictyostelium discoideum (slime mold),
  171–72
Didinium sp., 14
differentiation, 94. *See also* cell differentiation
diffusion. *See* thermal diffusion
digital mode, 82–84
DNA: base pairing, 146–50; nucleotide
  bases, 83–84; and protein structure, 62;
  and RNA, 106; transcription factors
  and, 191
Dobzhansky, Theodosius, 151
dogs: awareness in, 133; Pavlov's experiments, 35, 216–17
Dscam, 158

Einstein, Albert, 54–55
electronic pets, 43
Elowitz, Michael, 182
embryonic development, 191–94
endothelial cells, 177
Enterococcus faecalis, 170
environment, response to: amoebae and,
  236–37, 239–40; as basic to cell survival, 164–66; and cell evolution,
  155–56; as inherited memory, 233–36
enzyme reactions, cascading, 73–74
enzymes: and chemical bonds, 56–57, 60,
  60; as computational elements, 79–82,
  81; kinases, 75–79, *76, 78;* phosphatases, 75–79, *76;* protein regulation,
  61–62; regulation of manufacture of,
  61–63; as switches, 65–70, *67*